Sustainable Landscape Planning and Natural Resources Management

IEREK Interdisciplinary Series for Sustainable Development

Sustainable Landscape Planning and Natural Resources Management (SLNR) is a peer-reviewed book series that presents groundbreaking research through the intersection of two main areas that of which are 'Sustainable Landscape Planning' and 'Natural Resources Management'. The series initially covers a wide range of academic disciplines in sustainability, including landscape ecology, green urbanism, urban regeneration, land restoration and renaturation, energy management, and nature conservation. Moreover, it also incorporates the latest developments and research in engineering topics that contribute to more sustainable cities, such as Geosciences, infrastructure, civil engineering, geology, mining, water, and natural resources conservation and management.

By showcasing methods for discovering, managing, and employing natural resources, the series aims to provide approaches that enable the responsible use of these resources. Ultimately, the SLNR series is committed to advancing the United Nations' Sustainable Development Goals for 2030.

The wide range of topics covered in this series are primarily related to sustainable development, involving durable infrastructure projects, civil engineering, urban regeneration, the built, sustainability, climate change mitigation, energy management, geosciences, geology, and the future of cities, all of which depend on the availability of natural resources and effective wealth management. The series also features conceptual and empirical contributions from various interrelated fields of earth science, with an emphasis on the era of new discovery. It focuses on providing practical solutions to ensure sustainable development and provides new case studies that offer concrete examples of how to manage natural resources effectively.

The series encompasses the best research papers from various international conferences organized by IEREK conferences, also accepts proposals from other international conferences, including interdisciplinary contributions. It exhibits innovative approaches and highlights how they can best support both economic and sustainable development through the responsible use of natural resources and global, local, and individual activities for the welfare of all societies on Earth.

The SLNR series is a valuable source for experts in government, international organizations, consultancies, industry, business sectors, and academics. Published in cooperation with IEREK, the Springer SLNR series displays important new research in various sectors, delivering valuable insights and practical solutions to the complex challenges of sustainable development for the readers.

Javid Ahmad Parray
Editor

Soil and Land Use Change

Springer

Editor
Javid Ahmad Parray
Department of Environmental Science,
Government Degree College Eidgah Srinagar
Cluster University Srinagar
Srinagar, Jammu and Kashmir, India

ISSN 2948-1910 ISSN 2948-1929 (electronic)
Sustainable Landscape Planning and Natural Resources Management
ISBN 978-3-031-93074-4 ISBN 978-3-031-93075-1 (eBook)
https://doi.org/10.1007/978-3-031-93075-1

This Springer imprint is published by the registered company Springer Nature Switzerland AG
The registered company address is: Gewerbestrasse 11, 6330 Cham, Switzerland

Contents

Soil and Land Use Change: An Overview and Historical Developments 1
Neha Chauhan, Surbhi Antarkar, and Garima Bartariya

Human Intervention and Impact on Land Use and Land Cover Changes 17
Sonali S. Meher, Shantaj M. Deshbhratar, Vijay K. Hile, and Suraj P. Gajbhiye

Biochar and Sustainable Soil Health 27
Mahavir Gosavi, Lal Sahab Yadav, and Ashish Jain

Sustainable Soil Management—Principles and Importance 37
T. Dhanalakshmi and J. Rajichellam

Soil Restoration Methods and Challenges 45
Ajay Bhalerao and Ashish Jain

Influence of Cropping System on Soil Properties 55
Ahmed Abdul Haleem Khan

Climate as a Driver of Biodiversity Change on Land 63
Menakshi Dhar, Raman Jasrotia, and Seema Langer

Influence of Cropping System on Soil Properties 73
Suganthi Varathiyan, Dhanalakshmi Thiyagarajan, Subakanmani Swaminathan, Kerenhap Jyotsna Sweety, and Saranya Harikumar

Ecological Restoration and Reclamation of Degraded Lands in Indonesia: Eco-Physiological and Anatomy as an Advance Approach 79
Abban Putri Fiqa, Titut Yulistyarini, Sugeng Budiharta, Fauziah, Trimanto, Lia Hapsari, Apriyono Rahadiantoro, Janis Damaiyani, Sri Een Hartatik, Helbert, Masfiro Lailati, Agusdin Dharma Fefirenta, Mustaid Siregar, Edi Mirmanto, Lindsay F. Banin, and Enny Widyati

Unveiling the Dynamic Role of Microbes in Soil Management for Sustainable Agriculture 99
Nagma Khan, Baby Tabassum, Mohammad Hashim, and Asma Hasan

Climate Change: Impacts on Forest Cover 115
Partha Chandra Mondal, Arkadeb Mukhopadhyay, Madhu Tippannanavar, Shreosi Biswas, and Shweta Meshram

were common among nomadic societies. These methods cleared forests for temporary farming but often led to soil degradation when continuously practiced. Ancient civilizations like Mesopotamia, Egypt, the Indus Valley, and China made significant advancements in soil management. The Mesopotamians developed complicated irrigation systems to support agriculture in the arid region between the Tigris and Euphrates rivers (Jacobsen & Adams, 1958). Similarly, the Egyptians relied on the annual flooding of the Nile River to enhance their soils with nutrient-rich silt (Butzer, 1976). However, over-irrigation in Mesopotamia led to soil salinization, reducing agricultural productivity and contributing to the civilization's decline (Postgate, 1994). Similarly, deforestation and soil erosion in ancient Greece and Rome had long-term impacts on land fertility and the stability of these empires (Thirgood, 1981).

1.2 Mid to Early Modern Periods

1.2.1 Agricultural Expansion

During the medieval period world went through a significant agricultural expansion. The clearing of forests for farmland, known as the Great Clearance, transformed the landscape. Innovations like crop rotation and the three-field system improved soil fertility and good yield of agriculture.

1.2.2 Colonization and Global Impact

The Age of Exploration colonization had impacts on global land use and soil management. European societies introduced new crops and agricultural practices to the Americas, Australia, and other colonized regions. This often involved the large-scale transformation of native ecosystems into monoculture plantations, leading degradation of soil, biodiversity loss, and alteration of hydrological cycles (Crosby, 2004).The introduction of European farming practices and crops such as wheat, sugarcane, and tobacco had very intense effects on soil health. The intensive cultivation of these crops often led to soil exhaustion and erosion. Additionally, the displacement of indigenous land management practices, which were often more receptive to local environmental conditions, exacerbated soil degradation (Mann, 2005).

1.3 Industrial Revolution to the Twentieth Century with Technological Advancements

The Industrial Revolution brought significant technological advancements that transformed land use and soil management. Mechanized agriculture, use of synthetic fertilizers, and pesticides allowed for vast increases in agricultural productivity. However, these developments also had negative consequences for soil health. The very much use of chemical inputs often led to soil contamination, nutrients loss and a down fall in soil related biodiversity (Hillel, 2005).

By the early **twentieth** century, the consequences of unsustainable land use practices became increasingly very clear. A report of the Dust Bowl held in U.S. (1930s) highlighted the drastic effects of soil erosion. The establishment of the Natural Resources Conservation Service marked a significant change toward recognizing the significance of soil management programs or schemes (Worster, 1979). Globally, similar movements emerged in the United Kingdom, the establishment of the Soil Association in 1946 prompted organic farming practices. These early conservation efforts laid the practices for modern sustainable land management practices, emphasizing the need to balance agricultural productivity with environmental health (Paull, 2011).

1.4 Late Twentieth Century to Present with Industrial Globalization

The late **twentieth** century saw rise of industrial agriculture and globalization, further promoting land use changes. Large-scale monocultures, extensive use of chemical inputs, and the conversion of forests, pastures and weed lands into agricultural farms became common. While these practices significantly increase food production, they also led to severe soil degradation, greenhouse gas emissions, and loss of soil organic matter (Tilman et al., 2002).

In contrast to these challenges, the late twentieth and early twenty-first centuries have seen a growing land management practices. Concepts of agro ecology, permaculture, and regenerative farming promote methods that increase soil health, biodiversity, and ecosystem services.

2 Historical Developments in Soil and Land Use Change

The history of land use change is a testament to the evolving relationship between human societies and environment. This chapter traces the key developments in soil management and land use practices from ancient civilizations to the present day, highlighting how these changes have shaped landscapes, influenced agricultural productivity, and impacted ecological systems.

2.1 Ancient Civilizations and Early Agricultural Practices

2.1.1 Mesopotamia: The Cradle of Agriculture

Mesopotamia, often called the "Cradle of Civilization," saw the earliest forms of organized agriculture around 10,000 BCE. The evolution of irrigation systems in the Fertile Crescent allowed for the cultivation of wheat and barley, transforming migrant hunter-gatherer societies into settled agricultural communities (Fig. 1). However, rigorous irrigation led to soil salinization, reducing agricultural productivity over the time and contributing to the decline of these early civilizations (Jacobsen & Adams, 1958).

2.1.2 Ancient Egypt: Nile Valley Agriculture

In ancient Egypt, the random floods that reach every year in the Nile River deposited nutrient-rich silt on the flood leaving plains, leading to highly fertile soils ideal for agriculture. The Egyptians developed innovatory irrigation to manage the floodwaters, ensuring consistent crop yields (Fig. 1). This agricultural stability supported the growth of one of the most experienced civilizations in history (Butzer, 1976).

2.1.3 The Maya Civilization: Slash-And-Burn Agriculture

The Maya civilization in Mesoamerica used slash-and-burn agriculture, a method where forests were cleared and burned to create arable land (Fig. 1). While this technique allowed for short-term agricultural productivity, it led to long-term soil degradation and deforestation, contributing to the eventual down fall of the Maya civilization (Beach et al., 2006).

2.2 Medieval and Early Modern Developments

In medieval Europe, the feudal system shaped land use and agricultural practices. The open-field system, characterized by large, unfenced fields divided into strips and worked by farmers, dominated the landscape. By applying three-field crop rotation system, where one of the field remain fallow while the other two were planted with different crops, improved soil fertility and increased agricultural output and result (Campbell, 2016).

Fig. 1 Early agricultural practices

2.2.1 The Agricultural Revolution: Innovations and Escalation

The Agricultural Revolution of the eighteenth and nineteenth centuries marked a remarkable turning point in land use management. Innovations such as the seed drill, invented by Jethro Tull, and the introduction of new plant varieties and livestock breeds, transformed agricultural practices. The widespread practice of crop rotation, selective breeding, and the use of fertilizers caused a drastic increase in agricultural yield and reshaped rural landscapes (Overton, 1996).

2.2.2 Industrialization and Urbanization

The Industrial Revolution brought changes in land use through the mechanization of agriculture. Tractors, combine harvesters, and other machinery enabled large-scale farming, reducing manual labor and increasing output. The universal use of chemical fertilizers and insecticides further boosted crop yields but also led to degradation, pollution, and the loss of soil biodiversity (Matson et al., 1997).

2.2.3 Urbanization: Land Conversion and Soil Sealing

Expeditious urbanization during the nineteenth and twentieth centuries resulted in significant land use changes, with large areas of agricultural and natural land being converted into urban environments. Soil sealing, the veiling of soil surface with impermeable materials such as asphalt and concrete, disrupted natural soil functions, leading to increased runoff, reduced groundwater recharge, and the loss of cultivable land (Douglas, 1994).

2.2.4 Contemporary Issues and Sustainable Organic Practices

The Green Revolution of the mid-**twentieth** century introduced advanced irrigation techniques, and increased use of synthetic fertilizers to achieve high-yield crop varieties, these innovations increased crop production, particularly in developing countries, but also increased soil degradation, water scarcity, and environmental pollution (Evenson & Gollin, 2003). In response to the environmental challenges constitute by industrial agriculture, sustainable farming practices have gained fame. Conservation agriculture, which includes no-till farming, aims to maintain soil health, reduce erosion, and enhance biodiversity.

2.2.5 Land Restoration and Reforestation

Efforts to conflict land degradation and restore degraded soils have led to large-scale reforestation and afforestation projects (Fig. 2). Programs such as the Great Green Wall in Africa aim to combat desertification and improve soil health enhancing properties by planting trees and vegetation. These initiatives increased carbon sequestration, improve water retention, and restore ecological balance (Sessay et al., 2010).

3 Impact on Soil Properties

Land use change, the alteration of land from one usage to another, significantly impacts soil properties. These changes are driven by various human activities such as agriculture, urbanization, deforestation, and reforestation. Understanding the effects of these changes is crucial for soil conservation.

3.1 Soil Structure

Soil structure is made by arrangement of various soil particles into aggregate masses. It is crucial for soil health. Land use change can disturb basic soil structure by impacting the stability and size of soil aggregates. Agricultural practices, especially intensive tillage, often degrade soil structure, low porosity and increasing compaction. This compaction decreases water infiltration and root penetration, leading to very poor crop performance and increased erosion risks (Six et al., 2000).

3.2 Bulk Density

The mass of soil per unit volume is a key indicator of soil compression. It is termed as bulk density. Land use changes such as deforestation and urbanization typically increase bulk density due to the removal of vegetation and soil compaction from machinery and construction activities. Higher bulk density can limit root growth and water movement, effecting plant health and reducing agricultural productivity (Houghton, 1994).

3.3 Soil Organic Carbon

Soil Organic Carbon(SOC) is a crucial component of soil, effecting nutrient availability, soil structure and its fertility. Land use changes and its negative impact can lead to significant SOC losses due to reduction of organic matter inputs and increased decomposition rates. Conversely, practices like reforestation and conservation agriculture practices can enhance SOC levels by promoting organic matter accumulation and reducing erosion (Guo & Gifford, 2002).

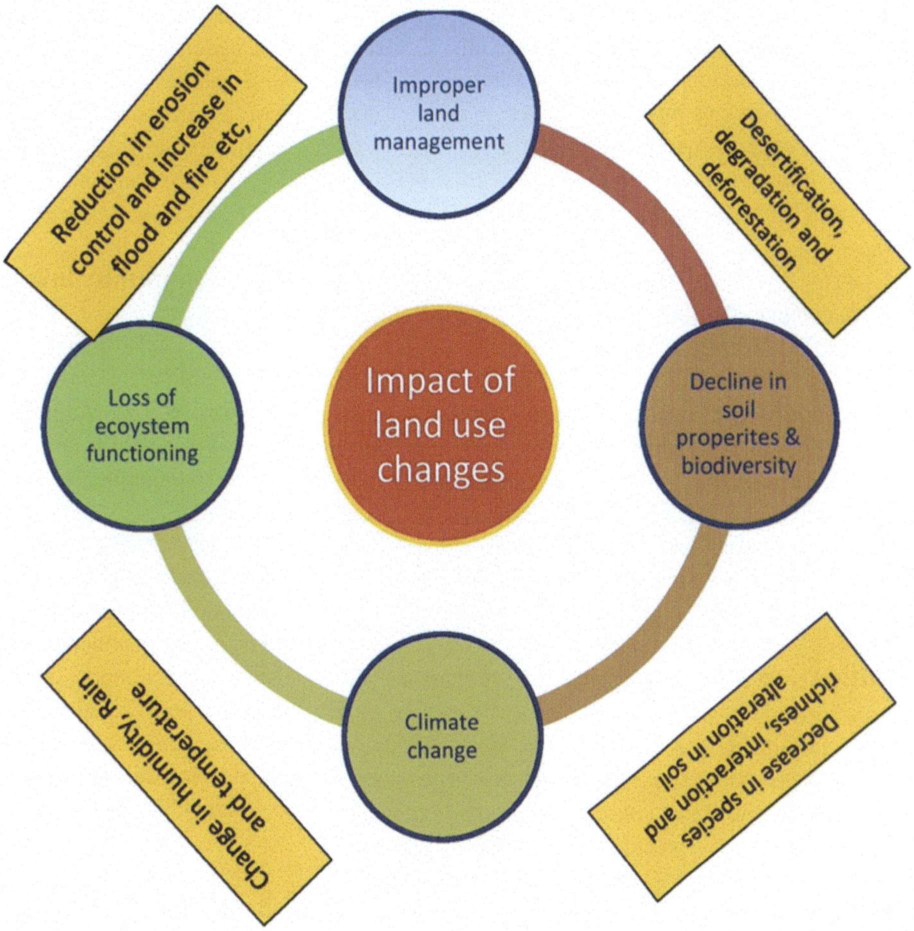

Fig. 2 Land use changes (Gamage et al., 2023)

3.4 Nutrient Availability

Nutrient availability, including essential elements like potassium nitrogen, phosphorus, is highly sensitive to land use changes. Agricultural intensification often leads to nutrient depletion due to continuous crop extraction and insufficient refilling through fertilizers. Additionally, urbanization can lead to nutrient imbalances caused by soil sealing and contamination from pollutants (Foley et al., 2005).

3.5 Soil Microbial Biomass

The living component organic matter of soil which is called biomass plays a vital role in nutrient cycling and soil structure maintenance. Land use changes can reduce microbial biomass due to habitat disruption and reduced organic matter inputs. Sustainable practices like organic farming and agroforestry can help maintain or enhance microbial biomass by preserving soil habitat and increasing organic inputs (Wardle, 1992).

3.6 Soil Biodiversity

Soil biodiversity, including bacteria, fungi, and invertebrates, is crucial for ecosystem functioning. Land use changes that degrade soil environments, such as urbanization and intensive agriculture, often lead to reductions in soil biodiversity. Conversely, diverse cropping systems and conservation practices can promote a rich soil biota, enhancing soil health and resilience (Altieri, 1999).

3.7 Case Studies

3.7.1 Deforestation in the Amazon

The Amazon rainforest is a prime example of impacts of extensive land uses on soil properties. Deforestation for agriculture and cattle ranching has led to significant soil degradation, including loss of soil carbon, reduced nutrient availability, and increased erosion. These changes have profound implications for local ecosystems and global carbon cycles (Fearnside, 2005).

3.7.2 Urbanization in China

Rapid urbanization in China has dramatically altered soil properties. Soil sealing in which the covering of soil occurs with impervious materials, has led to reduced water infiltration, increased runoff, and loss of arable land. Urban soils also face contamination from industrial pollutants, posing risks to organisms and environment (Chen et al., 2014).

3.7.3 Conservation Agriculture in Sub-Saharan Africa

Conservation agriculture, which includes practices like no-till farming, crop rotation, and cover cropping, has shown positive impacts on soil properties in Sub-Saharan Africa. These practices improve soil texture and basic structure, enhance SOC, and enhance nutrient availability, contributing to sustainable agricultural productivity and resilience against climate change (Giller et al., 2009).

4 Mitigation and Management Strategies

Adopting sustainable practices is essential to mitigate the negative impacts of land use change on soil properties. Techniques such as agroforestry, conservation tillage, and organic farming can enhance or protect health of soil by promoting biodiversity, improving its structure, and increasing SOC levels (Pretty, 2008).

4.1 Policy Implications

International initiatives and policies have a crucial role in promoting sustainable soil and its land use practices. Goal 15 (Life on Land) of the Sustainable Development Goals (SDGs), emphasize the importance of sustainable land management and soil conservation. The Global Soil Partnership, established by the Food and Agriculture Organization (FAO), aims to improve global soil governance and promote sustainable soil management (FAO, 2020).

The history of land use change is a testament to the intricate relationship between societies and the environment. From ancient irrigation systems to modern sustainable farming practices, the ways in which we manage soil and land have profound implications for environmental health and human well-being. Challenges of climate change, population growth, and environmental degradation, understanding the historical context of soil and land use change can inform more sustainable and resilient practices for the future. By learning from past successes and failures, we can work toward a more sustainable and equitable use of our planet's soil resources.

Effective policies and regulations are critical for managing land use changes. Governments can implement land use planning and zoning laws to control deforestation and urban sprawl. Incentives for sustainable farming practices and reforestation projects can also preserve soil health and preventing degradation (Lambin et al., 2001).

4.2 Community Involvement and Education

Engaging local communities and educating land users about the importance of soil conservation can lead to better land management practices. Participatory approaches of farmers and stakeholders in decision-making processes ensure that conservation strategies are practical and widely adopted (Reed, 2008).

Land use change significantly impacts soil properties, with consequences for agricultural productivity, environmental health, and climate change mitigation. Understanding these impacts and implementing sustainable management practices are crucial for maintaining soil health and ensuring long-term ecological balance. Through a combination of scientific research, policy intervention, and community engagement, the adverse effects of land use change on soil properties can be effectively mitigated.

4.3 Future Directions

Climate change poses new challenges for soil and land use management. Soils play a critical role in organic carbon sequestration, and practices that enhance carbon (SOC) levels in soil, such as agroforestry and regenerative agriculture, are gaining attention for their potential to mitigate climate change. Developing resilient agricultural systems that can adapt to changing climatic conditions is essential for future food security and environmental sustainability (Lal, 2004).

4.4 Technological Innovations and Precision Agriculture

Advancements in technology offer new tools for optimizing land use and soil management. These technologies enable farmers to look after soil health, manage inputs more efficiently, and reduce environmental impacts. Precision agriculture uses data-driven approaches to tailor farming practices in specific field conditions and promises to enhance productivity while conserving natural resources. (Gebbers & Adamchuk, 2010).

The history of soil and land use change reflects humanity's evolving relationship with the environment. From ancient agricultural practices to modern technological innovations, the ways in which we manage land and soil have profound implications for ecological health, food security, and sustainable development. As we face the challenges of the **twenty-first** century, adopting sustainable land use practices and leveraging technological advancements will be crucial for maintaining soil health and ensuring a resilient future.

5 Impact of Soil Alteration on Land Type

Food security, which is largely dependent on the impact of land used for it and agricultural practices of managing soil impact its quality, sustainability, and the safety of food produced from it. It has continued to be a global concern in response to growing human population demand. One of the most visible indicators of human activity contributing to land degradation is land use. The natural vegetation is removed in deforestation ether from overgrazing or from industrial emissions, which needs to be recovered in the form of implementing afforestation, the adoption of conservation agriculture, and the practice of agricultural methods with erosion control measures (Ojo et al., 2022).

Major changes that occur in land use are the component of the worldwide changes, which impact ecosystem structures, processes, and functions. The impact of changes in agricultural land use on composition of soil microbial communities and biomass continues to be widespread debated. Changes in the structure and functioning of microbial communities can adversely or positively affect ecosystem processes, e.g., decomposition of plant litter or the availability of nutrients. These changes are mainly due to soil characteristics like pH, depth, moisture, and temperature that can affect microbial communities and their biomass composition (Zhang et al., 2016). Human activity and accompanying land use change are the main causes of rapid degradation. The process of soil erosion significantly impacts the cycling of nutrients and worldwide social circumstances. It is well recognized that deforestation, overgrazing, tillage, and inappropriate agricultural practices lead to accelerated soil erosion, harm the soil, as do their mechanics. Impacts may be serious, not only due to land degradation and loss of fertility but also through the presence of a wide range of offsite effects such as sedimentation, siltation, or eutrophication of water channels in addition to increased flooding.

5.1 Soil Change in Cropland

The functioning of soil ecosystems around the world is affected by climate change and intensive land use. By the century's end, global forecast indicates that temperatures will increase by as much as 2 °C, while changes in rainfall patterns will lead to more frequent and severe droughts. In parallel, the demand for raw materials and foodstuffs is on the rise leading to the intensification of farming practices in a limited global area that has already dramatically changed ecosystems (Hirata et al., 2024). Land use leads to soil degradation in the tropics, where a mix of weather patterns and accelerated decomposition and nutrient loss are caused by poor soil management. The management practices determine soil quality which supports plant life and helps in maintaining or improving water and air quality. Therefore to ensure the sustainability of ecosystems, it is necessary to preserve and promote the quality of soil (Delelegn et al., 2017).

Soil Organic Carbon (SOC) represents one of the world's biggest land carbon pools, which can give significant feedback to atmospheric CO_2 concentrations through a slight change in magnitude. Suggestive indications for food security and soil health could be achieved by increasing levels of Organic C in cropland soils, as well as promoting carbon sequestration. The carbon inputs balance comes from residues and outputs due to microbial respiration is a primary determining factor for SOC storage on farmland. For example, crop productivity is reduced because of changing climatic factors, like temperature, precipitation, and heat, which may lead to a net loss of SOC (Ren et al., 2020).

Over the past decades, several approaches have been used to study world crop production in terms of inventory-based methods. In agricultural SOCs, global size and patterns vary due to significant variations in soil surface heterogeneity, intricate mechanisms, and a multitude of effect factors. Intensive agriculture methods are linked to additional detrimental environmental effects such increased nutrient leaching into rivers and pose a significant danger to SOC sequestration. Such type of discrepancies can be hurdle for global achievement of the 4.0 target. Agriculturists have been attempting to switch from intense management to recommended management practices (RMPs), which are more sustainable. Three key approaches have been identified, although the RMPs may vary from farm to farm, i.e., (1) organic amendments; (2) organic retention of wheat, rice straws and other straws (3) decreased tillage intensity. Recent studies have shown that RMPs can, in certain situations, increase SOC stocks within agro-ecosystems to a level like the land. Therefore, more soil carbon is to be sequestered in agricultural systems. (Lin et al., 2023) The Sustainable Development Goals of the United Nations for 2030 aim for striking a balance between conserving environmental services and raising agricultural productivity. In order to improve food security, preserve terrestrial ecosystems and fresh water, and slow down climate change. The 2030 Agenda for Sustainable Development advocates for national policies and international cooperation. These measures depend on standardized, independent, and timely data on agricultural extent and productivity. The 2019 Global Cropland Map states that rather than following national borders, global distribution and dynamics are a reflection of past land usage, population expansion, and

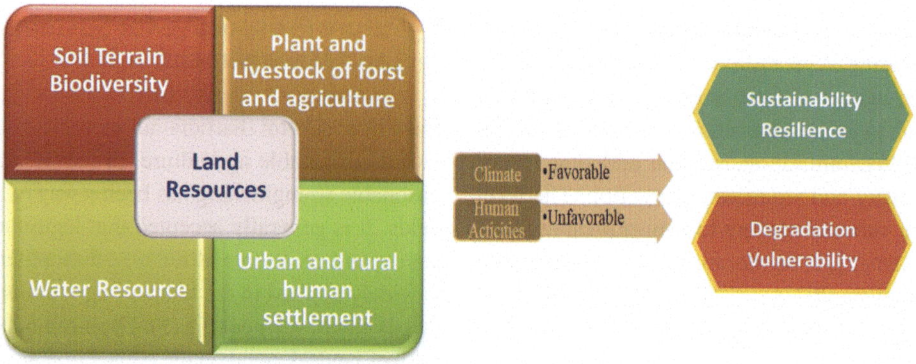

Fig. 3 Land resources (FAO)

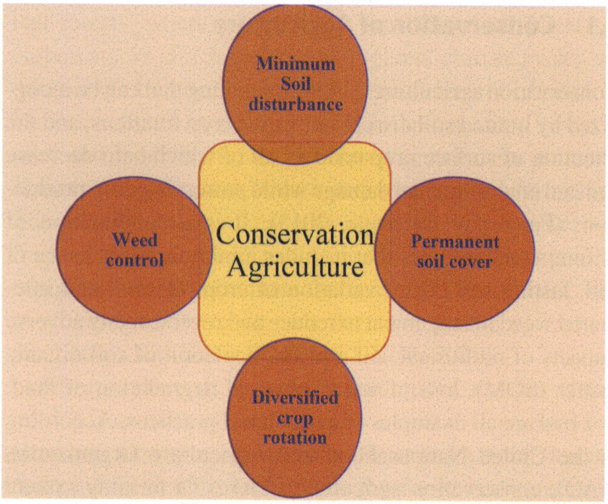

Fig. 4 Ways for conservation of agriculture (Farooq & Siddique, 2015)

minimal tillage. This method uses seed planting in tilled soil, leaving the preceding year's crop residue undamaged. It has been shown that the use of Zero tillage allows wheat to be planted in time and provides several advantages over conventional tillage methods. For ex, Zero tillage allows early sowing and decrease competition with Phalaris which is small, a herbicide-resistant weed capable of affecting wheat growth and yield in cases where wheat crops are planted late Conventional tillage. (Dev et al., 2023) Zero tillage (ZT) can lead to better soil quality by enhancing soil fertility, soil structure, and soil biological properties.

(ii) **Happy Seeder**: With a load of 10–12 tonnes, it is capable of sowing seeds and fertilizers in one single action, the soil is not tilled or disturbed and is capable of working with any type of crop residue. However, to operate effectively, it requires a double-clutch tractor with a maximum force of 400–500 rpm. The machine is equipped with nine times and weighs approximately 740 kg, which can be used in combination with a multi-crop planter box to cover many crops. To prevent seeds and fertilizers from falling at the correct depth in all nine rows, it is useful to use a depth controller. As per the studies, the utilization of the Happy Seeder method has the potential to yield an average of 20% more profits as opposed to farmers who choose to burn their crop residues. It was found that, compared to traditional farming practices, using the Happy Seeder for sowing led to a typical rise in yield about 10% in farmers' fields. When compared to zero till drilling, Happy Seeder could take a little longer to drill the hole. The use of the Happy Seeder can result in saving up to 45 L of fuel per hectare based on an average 35 HP tractor's fuel consumption rate of 6 L per hour (Dev et al., 2023).

(iii) **Laser Land Leveler**: The leveling of fields in India has been transformed by advanced computer-based technology, is laser land leveler, also known as "Computer Manjha" locally. To reduce the difference of elevation by less than 10 cm and lead to alterations in the production of crops and waterlogging in lower-lying areas, conventional land leveling methods were not able to do so. The field is automatically leveled with the laser land leveler, which reduces variability to a maximum of 2 cm.

A study was conducted to assess the effect of laser land elevation on nitrogen utilization efficiency in rice-wheat fields. Study findings showed that compared to areas treated with traditional methods, fertilizer use efficiency increased significantly in fields where laser levels were administered. Furthermore, considerable improvements in agricultural efficiency were seen in a rice and wheat crop system where the crops are cultivated at laser level (Dev et al., 2023).

(iv) **Multi-crop zero tillage**: A multi-crop planter is an agricultural tool designed for the efficient planting of multiple crops efficiently in one go is the main purpose of this machine, which is made up of three boxes. The initial box is specifically designed to store fertilizers, the second box is used to store seeds, and the third box is used to plant specific, maize and other crops that need accurate planting and have specific requirements for row spacing, depth, and power transmission., these planters are capable of handling variable seed sizes, rates, depths, and spacing for different crops. They are also equipped with a precise seed metering system, which allows planting various crops directly with accurate seed rate and spacing is achievable with the same fluted planter, unlike roller metering drills. This is made possible by using inclined rotary plates with adjustable groove numbers and sizes to accommodate different seed sizes and spacing requirements.

6.1.2 Practices Followed in Conservation Agriculture

- **Crop Diversification and Intensification**: One of the practices, of sowing different crops alternatively on the same piece of land while crop intensification is growing two to three crops in a sequence in the same year instead of doing one crop.
- **Nutrient management**: Farmers may use tools like leaf color charts to control the level of nutrients in their crops. A leaf color chart, which is a scale with five or six green stripes, is the first. The color of these strips changes from yellowish green to a deep green. With this chart, color of the leaves can be compared. If six out of ten leaves have some yellowish color and the farm's greenness score is less than four, then 25 kg of nitrogen per acre can be applied. To determine the nitrogen status of the plants, the color of the leaf chart is typically employed at predetermined intervals of 7 to 14 days.
- **Management of Crop Residue**: Around 100 million tons of crop residues are burned by farmers, leading to problems with the environment like greenhouse gas emissions, soil nutrient depletions and air pollution. Farmers are allowed to employ tractors like straw-handling happy seeders, zero tillage machines, paddy choppers, reaper combiners, and systems. and mulching equipment to manage crop residue. Crop residue can also produce animal feed, mushrooms, bio-fertilizers, mulching, ethanol, vermincomposting, and energy sources (Dev et al., 2023).

6.2 Organic Farming

Stabilizing agricultural production make it viable and feasible. A better way is needed because of the ever-increasing population, while at the same time, living resources such as food and water are becoming scarcer. The benefits of the green revolution credited to Dr. Swaminathan are now reaching a plateau and alternative techniques have become necessary given diminishing returns. Moreover, there is a problem of pollution caused due to the excessive use of fertilizers and artificial growth regulators. A natural balance between life and property for survival is currently crucial. Lord Northbound coined the term 'organic farming' in 1940. It was at the beginning of the 1800s that organic farming began. Justus Von Liebig developed a mineral plant nutrition theory in 1840. Some mineral salts might be a direct substitute for manure. India in the 2019–2020 period, approximately 2.75 million MT of Certified Organic Products, encompassing various food products, were manufactured, such as oilseeds, sugar cane, cereals and pulses, millets, aromatic & medicinal plants, fruits, coffee, tea, spices, dried fruit, vegetables, processed foodstuffs, etc. (Elayaraja & Vijai, 2020)

Currently, India is on the 9th rank in the world for organic agricultural land, which produces a variety of crops. The final product is achieved through the adoption of the concept and principles of Organic Agriculture. Organically grown foods are free from pesticides, environmentally friendly, lower the risk of chronic diseases, and safer than conventional products (Fig. 5). Organic products are grown in manure and compost, weeds, pests, and diseases can be controlled by naturally. In contrast, animals are organically fed & reared without the use of hormones or GMOs. Organic production in India is managed by two management bodies, namely the National Programme on Organic Production and Participatory Guarantees System-India (PGS-India). Organic certification covers an area of 3.566 million hectares, which includes cultivation and wild farming area in 2018–2019 (APEDA, 2019) and among all the states. Madhya Pradesh has the highest area under organic certification. The organic food market in India is growing at the highest rate of all, followed by clothing, beauty, and personal care. (Sandhya Kumari & Raj, 2019).

A compound annual growth rate of more than 23% is projected for the market, According to a report by Tech Sci Research Pvt Ltd, there will be an increase in organic cultivation area by 2023, supported by Governmental policies. The current Indian market is estimated at INR 40 000 million 2019 (APEDA, 2019). An organic substance is one that is not "artificial" or "synthetic," has a natural origin, or is created in a way that has no effect on any living species within the ecosystem (Fig. 6). In cities, there is a high demand for organically produced products. In this area, the following trends are developing: increasing demand of organically produced food, expanding online marketing platforms, increased creative investment, and entrepreneurial activity (Sandhya Kumari & Raj, 2019).

To increase and stabilize production, the majority of farmers are also interested in organically grown crops. With public support and market demand, awareness of food safety and environmental issues among consumers can be increased. They are increasingly concerned about their health and the environment. Consumers have come to view organic food as a meaningful option.

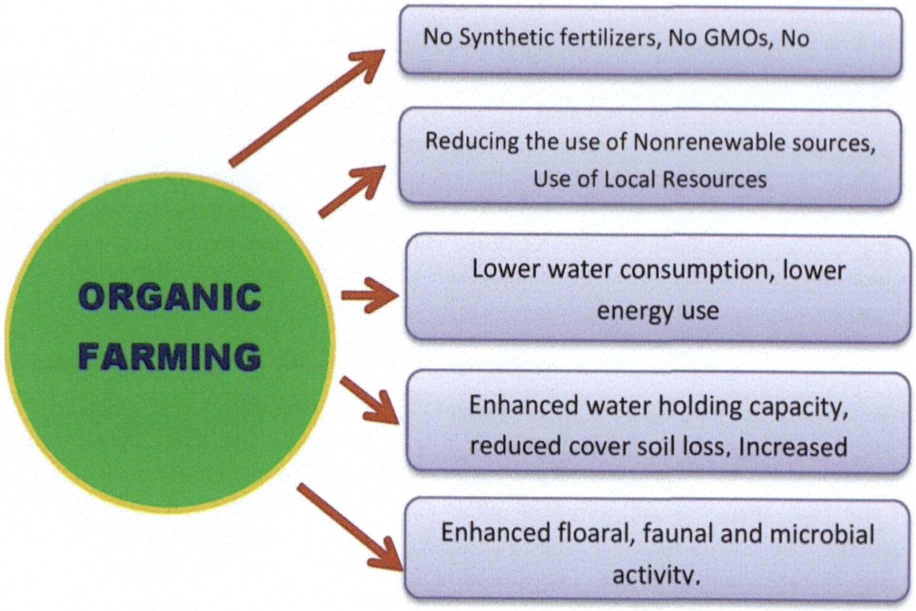

Fig. 5 Benefits of organic farming (Gamage et al., 2023)

Fig. 6 Comparison between organic and conventional foods (Sandhya Kumari & Raj, 2019)

Organic food Products	Conventional Food Products
• Cultivated using organic fertilizers such as manure and compost • Crop rotation, hand weeding, mulching, and tillage are used to manage weeds. • Pests controlled using bird, traps, and naturally derived pesticides	• Cultivated using synthetic and chemical fertilizers • Weeds managed by synthetic herbicides • Synthetic pesticides are used to control pests.
Organic meat, dairy and egg	Conventional meat, dairy and egg
• All organic feed is given to livestock so that they are free from hormones and GMOs. • Preventing disease involves using natural methods such as maintaining clean housing, practicing rotational grazing, and ensuring a healthy diet. • Outdoor access is provided for livestock.	• All livestock receive organic feed and are free from hormones and GMOs. • Livestock disease is prevented by administering antibiotics and medications .• Livestock might have access to the outdoors or they might not.

Human Health: The production and processing of Organic food is regulated by safety standards and regulations which confirms a very low concentration of chemicals and hazards.

GMO and Toxin-free food: Organic food and producers undergo very strict inspection and certification by third-party inspector which ensures the final quality of the product. No antibiotic or hormone is allowed to be given to the livestock.

High Nutrient Content: Organic fruits, vegetables, grains, and edible products are free from residues and pesticides, compared to traditional food to make them safer for consumption and to ensure that they contain enough nutrients.

Economics: The market price of certified organic food products is higher than the market price of traditional food items and straight away purchased by the company. It helps those farmers who have a supply chain system that is well-organized.

Environment: A natural method, such as mulching and crop rotations should apply. Green gas emissions reduction is achieved through residue management and the implementation of other enhanced organic practices., which helps mitigate climate change. (Sandhya Kumari & Raj, 2019)

6.3 Agroforestry

The system of agroforestry integrates traditional wisdom with contemporary scientific knowledge and innovative small-scale concepts. In doing so, they increase the possibilities for economic return of goods. Climate change is a major concern, and the solution is to adopt an agroforestry system that increases the resilience of the agricultural system to the sole crop growers, which will increase the forest and tree cover. Agroforestry involves integrating different land use systems that integrate trees and shrubs in agricultural areas improving productivity, profitability, and diversity can be achieved with or without the presence of livestock. as well as ecosystem sustainability according to the National Agroforestry Policy (2014).

Local people have been practicing traditional agroforestry practices since ancient times, deliberately using different multipurpose trees on their farms to maintain their density, frequency, and wide variety because of local and climatic conditions. For example, the farmers found *Acacia nilotica* to be highly beneficial due to its ability to absorb atmospheric nitrogen and its easy proliferation in the fields without any issues. Likewise, the widespread cultivation of *Azadirachta indica* has helped to mitigate deforestation, improve soil conditions, eliminate desertification areas, and restore world environmental temperature (Singh et al., 2021).

To tackle worldwide climate change and offer additional environmental and economic support, as well as social benefits, Agroforestry has risen to the forefront of land use strategy. Nevertheless, there is still a lack of systematic knowledge on the human and environmental effects of agroforestry practice and intervention.

Potentials of Agroforestry:

- Using the correct model for strengthening the productivity of the land, studying the interaction has been crucial for understanding the significance of the agroforestry system. and negative interactions help to increase productivity and decrease it. The type of model, the selection of species, the composition of varieties, and the nature of the family are important for the tree and the crop.
- The beneficial microorganisms and soil nutrient content have grown as a result of the trees' improved ability to absorb water through their roots. Agricultural forestry systems are good for the environment because they increase soil quality and lessen soil salinization and acidity. In the southern Indian state of Palani Hills, the usage of invasive leguminous trees, specifically acacia spp., has improved farming conditions for farmers by preventing unfavorable plant invasions into productive agroforestry systems.
- The agroforestry system's increased productivity and the limitation of land are both important factors to consider. In the same area of the field, agroforestry has a significant social impact on marginal and small communities' holdings of farmers.
- An ideal solution is, an agroforestry system to reduce the level of O_2 because it provides for the planting of trees on agricultural land and landscape. The agroforestry plantation should be established in a large size, and efforts to reduce greenhouse gases must be undertaken swiftly.
- By increasing the area of natural habitats in a limited area, such as the agricultural field, between existing reserves and the landscape, and by creating corridors, agroforestry contributes to reducing environmental pollution by planting landscape plantations with higher oxygen-producing plants, (Grigal, 2000)

7 Land Use Changes in Relation to Economic, Social, and Environmental Status

Land use transitions and rapid urbanization triggered many environmental changes and altered economic growth. Land use modifications are complicated issues that led to wide-ranging effects such as population expansion, infrastructure needs, and economic development. These modifications have negative as well as positive effects on economy, society, and environment.

7.1 Impact of Land Use Changes on Social Status

Community structure and human interaction are greatly influenced by land usage. It also effect on social fabric of communities. Large-scale land use changes, especially for infrastructure and urban expansion become a main cause displacement of communities that results loss of cultural heritage and livelihoods. Schmink and Wood (1992) stated that in the Amazon rainforests are facing threats posed by logging and deforestation that effects on indigenous tribes who are especially sensitive for their heritage and agricultural land usage. Day by day urban development leads to conversion of farms and forests in concrete building and reducing food and lumber producing forests. These intensive agriculture practices degrade quality of land resource outputs. Furthermore these land management strategies are frequently implemented with the intention of achieving short-term goals to raising crop yield. However, these changes cause disturbance of forest ecosystems and leave long-term consequences on the supply of ecosystem, such as flood mitigation, erosion prevention, and climate regulation, (Weiss et al., 2020).

Land use shifts from traditional farming to commercial agriculture affected in rural communities. These transformations provide new job opportunities to certain high class people while others of lower classes can lose have to face loss of livelihoods and food insecurity as well as their cultural significances. This is evident in Southeast Asia where the transformation of paddy fields into aquaculture had an effect on social dynamics and local food systems (Hall et al., 2011). Social inequality gap increased by result of urbanization and give rise to slums or poor shanty towns. There areas have to face serious public health issues because of lacking essential amenities like clean water, sanitary conditions, and healthcare. Urban poor people are more vulnerable to environmental risks and social marginalization. A decrease in rainfall in the southern regions also prompted a transition from grazing to dry-land crops, among other reasons driving land use change. Variations in climatic variability create unwanted changes in land covers that have impact on social life of communities.

7.2 Impact of Land Use Changes on Economic Status

Land use changes in agricultural and non-agricultural areas become worsen with increased urbanization. Therefore many local laws are made to pay those farmers who face negative effect of these transformation imposed on their livelihoods. Farmers' labor expenses may increase due to competition for workers from non-agricultural businesses. When all agricultural supporting sectors disappear and cropland falls below a certain mass, the local agricultural economy may collapse. Farmers can benefit greatly from urbanization as well. Farmers now have more chances to sell higher-value commodities because of the advent of a new consumer base. For instance, vegetable growers typically obtain better prices in metropolitan regions (Lopez et al., 1988). How quickly agricultural economies may change is demonstrated by the proliferation of vegetable farms, wineries, and other crop sectors in many regions. Considerable flexibility has been displayed by numerous farmers in modifying their businesses to capitalize on emerging economic prospects in the periphery of cities. In places with dense populations, they farm more intensively (Lockeretz, 1988).

Urbanization stimulates economic activity and creates jobs to improve infrastructure development. However, the reestablishment of forests and agricultural land into urban areas frequently leads to issues with housing affordability, urban sprawl, and increased property values. For example, rapid urbanization in China has caused social displacement and environmental degradation in addition to contributing significantly to the country's economic boom (Liu et al., 2010). Historically, economic expansion has been stimulated by the conversion of natural landscapes into mechanical agricultural farms. This shifting of land's traditional use to modern land utilization altered the economic status of communities as well as their interest in sustainable approaches. For example, millions of people were lifted out of poverty by the Green Revolution. Nonetheless, there may be drawbacks to agricultural expansion's financial advantages in long terms. Short-term increases in GDP and employment may be accompanied by decreased biodiversity, degraded soil, and decreased long-term agricultural output (Tilman et al., 2002). The patterns of land use are considerably changed by infrastructure developments including industrial zones, dams, and roadways. By supplying electricity, promoting trade, and enhancing connection, these projects can promote economic development. But, these projects may additionally result in the deterioration of crop land and the uprooting of local communities. In case of economic growth, urbanization, and industrialization can lead to the creation of new jobs in the building, real estate, and service sectors. Farming lands are converted to more productive use, like intensive agriculture. These may boost food supply to keep up with rising demand. People may benefit from possibilities and increased economic growth as a result through strategic planning. Incentivizing farmers to adopt sustainable practices can help preserve biodiversity and soil health. Absorption of farm lands by urban development might result in a decrease in food production and cause a possible increase in food prices. These changes in land usage may result in biodiversity loss and deforestation. In the long run, these environmental issues may result in decreased crop yields, more flooding, and poorer water quality, among other economic consequences. Market forces don't always account for all of the environmental costs associated with a given land use. This may result in the undervaluation of natural resources and ineffective land use planning.

7.3 Impact of Land Use Changes on Environmental Status

Changes in land cover have a significant negative influence on the climate. Uncontrolled land use led by agricultural and urban growth also degrades ecosystems and the environment. Natural environments like wetlands and forests provide essential habitat to many species. When these natural habitats are converted into urban development or agriculture, it threatens biodiversity. Although the fragmentation and destruction of wildlife habitat occurs due to these land use change for betterment of human life. It leads to a huge biodiversity loss. This decline in biodiversity occurs due to deforestation for logging, agriculture, and urban growth. Laurance et al. (2014) stated that tropical rainforests, which are home to a large number of species, faced such type of threats. The Amazon region has witnessed massive deforestation, which causes extinction of species, loss of habitat, and disturbances to ecosystem services.

Soil erosion and desertification may come from practices like intensive agriculture and urbanization. As a result, the land's productivity and capacity to store carbon are decreased. Desertification and soil deterioration can result from excessive grazing, monoculture and inappropriate irrigation. A report of UNCCD (2017) published about unsustainable land use practices in parts of Africa like the Sahel. These practices leads to desertification and exacerbated endangering livelihoods and food security.

Forests traps greenhouse gases like carbon dioxide and maintain a balanced temperature in atmosphere. But deforestation cause Global warming by releasing this stored carbon into the atmosphere. The original land covers which degraded by alteration in natural land habitat also change local weather patterns.

Water quality can be adversely affected by changes in land use. Vegetation loss is considered as a major reason of sedimentation in streams and rivers and soil erosion. Furthermore, run off of agricultural pesticides and fertilizers used in agricultural practices also becoming a severe reason to contaminate or poison water bodies. Increased surface runoff from urbanization lowers groundwater recharge and raises the danger of flooding. Droughts, landslides, and floods can become more perhaps as a result of the loss and deforestation of wetlands. The infrastructure and human population may suffer greatly as a result of these natural calamities.

References

Agricultural and Processed Food Products Export Development Authority (APEDA). (2019). *National programme for organic production*. Retrieved from apeda.gov.in.

Ahirwal, J., Kumari, S., Singh, A. K., Kumar, A., & Maiti, S. K. (2020). Changes in soil properties and carbon fluxes following afforestation and agriculture in tropical forests. *Ecological Indicators*, 123.

Altieri, M. A. (1999). The ecological role of biodiversity in agroecosystems. *Agriculture, Ecosystems & Environment, 74*(1–3), 19–31.

Beach, T., Luzzadder-Beach, S., & Dunning, N. (2006). Ancient maya impacts on the earth's surface: An early anthropocene analog? *Quaternary Science Reviews, 25*(13–14), 1441–1462.

Butzer, K. W. (1976). *Early hydraulic civilization in Egypt: a study in cultural ecology*. University of Chicago Press.

Campbell, B. M. S. (2016). *The great transition: Climate disease and society in the late-medieval world*. Cambridge University Press World.

Chen, M., Lu, W., & Chang, C. (2014). A study on the impact of urbanization on soil properties in China. *Environmental Science and Pollution Research, 21*(2), 793–801.

Clunes, J., Valle, S., Dörner, J., Campos, M., Medina, J., Zuern, S., & Lagos, L. (2022). Changes in soil quality of an urban wetland as a result of anthropogenic disturbance. *Land, 11*(3), 394.

Crosby, A. W. (2004). *Ecological imperialism: The biological expansion of Europe, 900–1900*. Cambridge University Press.

Delelegn, Y. T., Purahong, W., & Blazevic, A. (2017) Changes in land use alter soil quality and aggregate stability in the highlands of northern Ethiopia. *Scientific Report, 7*, 1360

Dev, P., Khandelwal, S., Yadav, S., Arya, V., Mali, H., Dr, P., & Yadav, K. (2023). Conservation agriculture for sustainable agriculture. *International Journal of Plant & Soil Science., 35*, 1–11.

Donovan, M., & Monaghan, R. (2021). Impacts of grazing on ground cover, soil physical properties and soil loss via surface erosion: A novel geospatial modelling approach. *Journal of Environmental Management, 287*, 112206.

Douglas, I. (1994). Human settlements. In W. B. Meyer & B. L. Turner (Eds.), *Changes in land use and land cover: A global perspective* (pp. 149–169). Cambridge University Press.

Elayaraja, M., & Vijai, C. (2020). Organic farming in India: Benefits and challenges. *European Journal of Molecular & Clinical Medicine, 7*(11), 3021–3029.

Elliot, W. J., Page-Dumroese, D., & Robichaud, P. R. (2018). The effects of forest management on erosion and soil productivity.

Evenson, R. E., & Gollin, D. (2003). Assessing the impact of the green revolution, 1960 to 2000. *Science, 300*(5620), 758–762.

Ewing, J. M., Vepraskas, M. J., Broome, S. W., & White, J. G. (2012). Changes in wetland soil morphological and chemical properties after 15, 20, and 30 years of agricultural production. *Geoderma, 179–180*, 73–80.

FAO. (2020). *Global soil partnership*. Retrieved from FAO website.

FAO and UNEP. (2020). *The state of the World's Forests 2020. Forests, biodiversity and people*. Rome.

the conversion of natural ecosystems into croplands, which is another important factor contributing to LULCC. These modifications may affect biodiversity, lower the availability of water, and deteriorate soil quality (Nuissl & Siedentop, 2021). An essential element of LULCC is deforestation, which is mostly caused by logging, agriculture, and infrastructural growth. Because deforestation releases carbon dioxide that is stored in trees, it can cause habitat loss, soil erosion, and changes in local and global temperatures (Foley et al., 2005). Roads, dams, and urban growth are examples of infrastructure development that can exacerbate LULCC by fracturing habitats, changing water flows, and upsetting natural ecosystems. The lives of humans, environmental services, and biodiversity may all be significantly impacted by these changes (Turner et al., 2007) It is important to research Land Use and Land Cover Changes (LULCC) in relation to human involvement for a number of reasons.

Environmental Impact: Human activities like agriculture, urbanization, and deforestation can drastically modify the environment, resulting in biodiversity loss, ecosystem alterations, and modifications to regional and global climate trends (Geist & Lambin, 2002). Resource Management: By guaranteeing the sustainable use of natural resources like water, soil, and forests for future generations, a greater understanding of LULCC aids in better resource management (Eslamian et al., 2024).

Policy Development: Understanding the results of LULCC studies can help policymakers create land-use plans and management plans that are effective in promoting conservation and sustainable development. Ecosystem Services: The provision of ecosystem services—such as clean water, air purification, and carbon sequestration—that are necessary for human well-being may be impacted by changes in land use and land cover (Prem & Sheikh, 2023). Social and Economic Impacts: LULCC may have a major impact on the distribution of wealth, livelihoods, and food security, especially in rural communities that rely heavily on natural resources.

Climate modify: Through their effects on local and regional climate patterns, land use changes like deforestation and urban growth can modify the balance of greenhouse gases in the atmosphere. Changes in land use and cover (LULCC) are caused by a complex interaction between human and natural forces. An outline of the main forces behind LULCC is provided below: Population Growth: As the world's population rises, so does the need for land for infrastructure, agriculture, and housing, which causes changes in the way that land is used (Rai et al., 2023).

Urbanization: The process of turning rural land into an urban region results in changes to the land cover, including the disappearance of agricultural and forest areas. Agricultural development: One of the main causes of LULCC is the development of agricultural land to accommodate the rising demand for food and biofuels, which results in the conversion of natural ecosystems and deforestation.

Infrastructure Development: Converting land is often necessary for the building of roads, highways, dams, and other infrastructure projects, which results in alterations to the land cover.

Industrialization: The conversion of natural habitats into industrial zones is one way that changes in land use and cover can result from industrial activity. Climate Change: By altering temperature, precipitation, and other environmental parameters, climate change can have an impact on LULCC and cause changes in patterns of land cover.

Policy and Governance: The scope and makeup of LULCC are significantly influenced by land use policies, land tenure schemes, and governance frameworks.

Technological Advancements: By enabling more effective land use practices and the monitoring of changes in land cover, technological advancements like mechanized agriculture and remote sensing can have an impact on LULCC.

Economic considerations: ***LULCC*** can be driven by and have an impact on land use decisions based on economic considerations such as land pricing and market demand for agricultural products.

Cultural and Social Factors: LULCC patterns can also be influenced by social norms, traditional land use systems, and cultural traditions.

2 Historical Perspective

There was a substantial change in the way land was used when settled agriculture replaced hunting and gathering. This resulted in land clearing for towns and deforestation, mostly for agricultural purposes. One of the main causes of LULCC has been the expansion of urbanization. Forests, pastures, and agricultural land are transformed into built-up regions as urban centers encroach on neighboring rural areas. As industries, infrastructure, and transportation networks were built, the Industrial Revolution significantly altered how land was used. As a result, there was an increase in urbanization and the transformation of rural areas into industrial zones (Ponting, 2007).

As a result of the necessity to feed a growing population, agricultural land has expanded. As a result, wetlands, meadows, and woods have been converted to farms, resulting in the loss of habitat and a fall in biodiversity. Natural landscapes have been altered as a result of the building of roads, highways, dams, and other infrastructure developments. Ecosystems have been disturbed and habitats divided as a result. Mostly due to logging, agriculture, and urbanization, deforestation has been a major cause of LULCC. Deforestation causes local climatic patterns to shift, habitat loss, and soil erosion. As a result of rising demand for goods like soy, cattle, and palm oil, land use in areas like Southeast Asia and the Amazon rainforest has changed. A number of causes, including shifting precipitation patterns, rising sea levels, and temperature rises, are causing changes in land cover patterns. Changes in land use and vegetation kinds may result from climate changes.

3 Major Events or Periods that Significantly Impacted LULCC

Over time, a variety of occasions and eras have had a substantial influence on Land Use and Land Cover Changes (LULCC). Commencing approximately 10,000 years ago, the Agricultural Revolution was one of the first significant occurrences. Around this time, farming replaced hunting and gathering, which resulted in extensive deforestation and the transformation of natural habitats into agricultural land. Beginning in the eighteenth century, the Industrial Revolution was another important era. During this time, there was a significant increase in the conversion of land for infrastructure, factories, and cities due to the fast-paced industrialization, urbanization, and growth of transportation networks.

The Green Revolution, which started in the 1940s, brought about another significant phase of transformation in the twentieth century. During this time, high-yield crop types, fertilizers, and pesticides were introduced (Evenson & Gollin, 2003). This improved agricultural production was accompanied by significant deforestation and land degradation. More recently, as a result of growing globalization and the demand for commodities, mining and agriculture have expanded, further degrading the land and converting it into new uses.

Human actions and events have had a substantial historical impact on the LULCC, resulting in patterns of land use and cover changes that have important environmental and socioeconomic ramifications.

4 Drivers of Change

Numerous factors, both natural and man-made, have an impact on Land Use and Land Cover Change (LULCC).

Population Growth: As the population grows, there is a greater demand for resources like food, energy, and housing, which may cause patterns of land use to shift. Economic Development: Land cover changes, such as deforestation, the conversion of natural habitats, and the growth of urban areas, are frequently the consequence of economic activity like agriculture, industry and urbanization.

Technological Development: Better farming methods, more effective mining methods, and better infrastructure construction are just a few examples of how technological advancements can affect how land is used (Geist & Lambin, 2002). Governance and Policy: Government rules and policies, such as zoning laws, land tenure schemes, and conservation policies, can affect how land is used.

Climate Change: Modifications in temperature and precipitation patterns can have a direct effect on land cover through changes in vegetation, and they can also have an indirect effect on agriculture and water resources.

Infrastructure Development: Building roads, dams, and other infrastructure may cause patterns of land use to shift, resulting in habitat fragmentation and a rise in urbanization ((Nuissl & Siedentop, 2021). Cultural and Socioeconomic Factors: Traditional farming methods and patterns of land inheritance are examples of how cultural ideas and socioeconomic circumstances can affect decisions about how to use land. These factors can affect the LULCC pattern in a number of ways, such as: Spatial Distribution: The effects of distinct factors on various regions may differ, resulting in spatially varied LULCC patterns.

Temporal Dynamics: Variations in seasonal agricultural land use are one example of how drivers can alter over time and cause temporal shifts in LULCC patterns. Significant ecological effects of LULCC may include habitat loss, fragmentation, and degradation, all of which may have an adverse effect on biodiversity and ecosystem services. Feedback Loops: Modifications to the land use may also have a feedback effect on the drivers, resulting in intricate feedback loops that further affect LULCC patterns.

5 Impact on Land Use

Changes in land use patterns have been mostly driven by human action, dramatically changing landscapes all over the world. In an effort to meet the world's rising food demands, agriculture has grown, resulting in habitat change, deforestation, and biodiversity loss. Vast swathes of land have been converted into towns and infrastructure as a result of urbanization, splintering ecosystems and decreasing the amount of land available for wildlife and agriculture. Mining and manufacturing are two examples of industrial operations that have further damaged the land and contaminated the water and soil.

which can exacerbate socioeconomic inequality by causing disputes over how best to exploit these resources. Economic Repercussions: Agricultural Productivity: In areas where agriculture is the main economic activity, changes in land use patterns may have an influence on agricultural productivity, which may have an impact on food security and livelihoods. The availability and quality of land resources, microclimates, and water availability are all impacted by Land Use and Land Cover Change (LULCC), which has a substantial impact on agricultural output. Ecosystem balance is upset when natural landscapes like wetlands, forests, or grasslands are transformed into agricultural land, cities, or industrial locations (Wang et al., 2023) This can result in decreased soil fertility, increased erosion, and deterioration of the soil, all of which make the area less suitable for growing crops. Additionally, by influencing temperature, humidity, and precipitation patterns, changes in land cover can alter local climates. For instance, a region's deforestation may see less rainfall and higher temperatures, which would make the environment less conducive to crop growth.

The growth of urban areas in many emerging nations, where productive agricultural land is progressively being converted into residential, commercial, and industrial zones, is a germane illustration of how LULCC affects agricultural production. Rapid urbanization has resulted in the loss of fertile land that was formerly used for agricultural in areas like the suburbs of Mumbai. This results in a decrease in food production as well as a reduction in the area that may be used for agriculture. Urban sprawl's loss of agricultural land pushes farmers to farm less productive land, which lowers yields and jeopardizes the region's food security.

One area in Maharashtra, India, where LULCC can have an effect on agriculture is the Thane region. In addition to reducing the amount of farmland available, the city's growth into agricultural areas has changed the local water regimes, creating a shortage of water for irrigation. Farmers may give up on traditional methods as agricultural output drops, which would further exacerbate the issue by resulting in the extinction of regional varieties and traditional knowledge. This exemplifies how LULCC can cascade into agricultural systems, diminishing food production's resilience and sustainability in the face of environmental and socioeconomic shifts.

The Thane region of Maharashtra, India gives an example of how LULCC might affect agriculture. There is now less farmland available due to the city's development into agricultural areas, and the local water regimes have changed as a result, making irrigation water scarce. Compounding the issue, farmers may give up on traditional methods when agricultural production drops, which would result in the extinction of regional variety and traditional knowledge. The resilience and sustainability of food production in the face of environmental and socioeconomic changes can be diminished as a result of the cascade effect that LULCC can have on agricultural systems.

Ecosystem Services: Pollination, water purification, and climate regulation are examples of ecosystem services that are impacted by LULCC and have a financial value but may diminish as a result of changes in land cover. Infrastructure Costs: Investing in roads, irrigation systems, and housing is often necessary when land use changes, and this can put a financial strain on local governments and communities.

8 Impact on Land Cover and Ecological Implications of These Changes

There are several ecological ramifications from the substantial changes in land cover types brought about by human activity. Among the significant changes are:

1. **Deforestation**: The removal of forests for logging, urbanization, and agriculture has resulted in increasing greenhouse gas emissions, ecosystem disruption, and biodiversity loss. Deforestation is greatly impacted by Land Use and Land Cover Changes (LULCC), which are frequently brought on by increased agricultural production, urbanization, and infrastructural development. Converting land that was once covered in forests to other uses results in the loss of biodiversity, disturbs ecosystems, and releases stored carbon, which fuels climate change. Large-scale forested area removal is a direct result of the demand for agricultural land in many places, which is directly linked to deforestation. This decreases the amount of forest area and also splits up the forest that is left, which makes it harder for species to survive and puts them at risk of going extinct (Geist & Lambin, 2002).

   An excellent illustration of how LULCC contributes to deforestation is seen in the Amazon jungle. Large areas of forest have been removed in Brazil for logging, soy farming, and cattle grazing. Known as the "lungs of the Earth," the Amazon is home to a wide variety of species and is essential in controlling the world's temperature. However, the region has seen a sharp decline in the amount of forest cover as a result of changes in land cover brought about by economic activity. Deforestation has a terrible impact on indigenous communities and biodiversity, increases greenhouse gas emissions, and modifies regional and global climate patterns. The state of the Amazon highlights the critical need for sustainable land-use policies that strike a balance between environmental preservation and economic growth.

2. **City Development**: Natural ecosystems have been transformed into built-up regions as a result of town and city growth, which has resulted in habitat loss, fragmentation, and isolation of wildlife populations. Land Use and Land Cover Change (LULCC) has a major influence on city growth and frequently modifies the urban environment dramatically (Muche et al., 2023) The conversion of natural landscapes into constructed environments has a major effect and can result in the loss of open areas, farmland, and natural habitats. Forests and wetlands are routinely destroyed to create room for new residential, business, and industrial constructions as cities grow. The ecosystem services that are essential for preserving the quality of the urban environment, like temperature regulation, carbon sequestration, and air and water purification, may decline as a result of this transition. Furthermore, because more building is removing permeable areas, floods may worsen because precipitation cannot naturally seep into the earth, increasing runoff and stressing urban drainage systems.

   An appropriate illustration of how LULCC has affected city development is Mumbai, India. Mumbai's land use patterns have changed significantly as a result of the city's increasing urbanization, mainly the conversion of mangrove forests and coastal wetlands into residential and commercial areas. In addition to upsetting the surrounding ecosystems, this has made the city more susceptible to the effects of climate change, such as rising sea levels and storm surges. Particularly during the monsoon season, some parts of the city are now more vulnerable to floods due to the loss of natural barriers like mangroves.
3. **Agricultural expansion**: The use of agrochemicals has resulted in habitat loss, fragmentation, and deterioration of soil and water quality as a result of the conversion of natural habitats into agricultural land.
4. **Mining**: mining operations have impacted biodiversity and ecosystem services by causing the destruction of trees, contaminating soil and water, and uprooting communities.
5. **Infrastructure development**: Building roads, dams, and other infrastructure projects has changed hydrological processes, fragmented habitats, and raised the possibility of natural disasters.

Numerous ecological ramifications result from these changes, including:

1. **Loss of biodiversity**: The loss of species and genetic variety brought about by habitat degradation and fragmentation has decreased ecosystem resilience and its capacity to adapt to changing environmental conditions.
2. **Changes in ecosystem services**: Modifications to the types of land cover have an effect on ecosystem services that are important to human health, including soil fertility, carbon sequestration, and water purification.
3. **Climate change**: By releasing stored carbon into the atmosphere and changing the Earth's albedo, deforestation and other land cover changes affect the planet's temperature and precipitation patterns.
4. **Water resources**: Land cover changes may have an effect on the quantity and quality of available water, which may have an effect on ecosystems and human populations who rely on these resources.
5. **Human well-being**: The loss of natural habitats, greater exposure to air and water pollution, and modifications to disease vectors and patterns are just a few ways that changes in land cover can have an impact on human health. All things considered, land cover types have been dramatically changed by human activity, with substantial ecological ramifications that emphasize the necessity of sustainable land use practices and conservation initiatives.

Examples of specific regions or ecosystems where human intervention has had a significant impact on LULCC.

There are numerous instances worldwide where changes in land use and land cover have been profoundly altered by human activity (LULCC). Here are some noteworthy examples: Brazil's Amazon Rainforest: The region's land cover has changed significantly as a result of massive logging, urbanization, and agricultural destruction. This has worldwide ramifications for climate change in addition to impacting the region's biodiversity. Australia's Great Barrier Reef: Due to human activities including pollution, overfishing, and climate change, the reef has been degraded and coral bleached, which has an impact on both the marine ecology and coastal land use.

The Aral Sea in Central Asia: Due to river diversion for irrigation, the Aral Sea's water levels have drastically decreased, which has had a substantial impact on the surrounding area's land usage and land cover, including desertification and biodiversity loss. The Mississippi River Delta in the United States has seen changes in wetland habitats and patterns of land use due to human alterations such as levees, dams, and channelization, which have affected the Mississippi River's natural flow. The Sundarbans, which are located in India and Bangladesh, have seen substantial changes in land use and land cover as a result of human activities including deforestation, urbanization, and shrimp farming (Shiva, 1991). These changes have had an effect on the mangrove ecosystem and biodiversity of the area (Fig. 2).

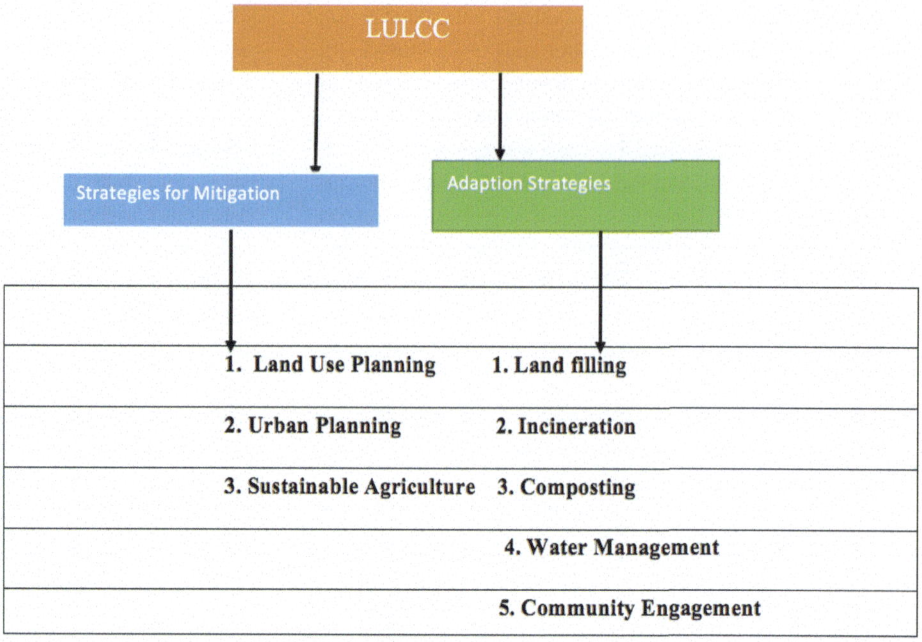

Fig. 2 Mitigation and adaptation for LULCC

9 Mitigation and Adaptation

9.1 Strategies for Mitigation

Land Use Planning: Putting protected areas, zoning laws, and land use plans into practice to control development and maintain natural ecosystems. Planting trees to absorb carbon dioxide and rebuild ecosystems is known as afforestation and reforestation.

Sustainable agriculture: In order to lessen the negative consequences of land use and land cover change (LULCC), sustainable agriculture is essential. LULCC frequently results in soil degradation, biodiversity loss, and deforestation but sustainable farming techniques can help mitigate these effects by supporting environmental protection and maintaining food security (Meadows et al., 1992) Crop rotation, conservation tillage, and agroforestry are a few practices that improve soil health, lessen erosion, and sequester carbon, all of which help to mitigate the effects of climate change. Agroforestry increases carbon storage and biodiversity by incorporating trees and shrubs into agricultural landscapes, resulting in a resilient ecosystem that can survive the effects of climate change.

Furthermore, by increasing the productivity of already-existing agricultural land, sustainable agriculture lessens the need for significant land conversion. Organic farming lowers pollution and preserves soil fertility by eschewing synthetic pesticides and fertilizers. By fostering diversity within farming systems, techniques like permaculture and polyculture increase the systems' resistance to pests and illnesses. By preventing additional expansion into natural ecosystems and preserving the ecological balance, these techniques help lessen the detrimental effects of LULCC. By implementing these tactics, sustainable agriculture contributes significantly to the preservation of natural resources and shields the ecosystem from the damaging impacts of changing land use. It also ensures the long-term viability of farming.

Urban planning is the process of creating smaller, more compact cities while minimizing the need for urban sprawl and protecting open spaces. Energy efficiency is the process of lowering energy consumption through effective methods and technologies, which in turn lessens the demand for energy production that requires a lot of land.

9.2 Adaptation Strategies

Restoring damaged landscapes to increase their resilience and capacity to deliver ecosystem services is known as ecosystem restoration.

Landfilling-This is putting the waste in a lined pit or mount with suitable leachate and landfill gas control, in compacted layers. Nowadays, there's a push to take advantage of landfills' microbial methanogenic activities and use them as bioreactors to get rid of garbage and use the methane produced as a substitute energy source.

Incineration-The process of incinerating garbage involves feeding the screened material through a chamber that is heated to a high temperature (around 1200 °C) and has enough oxygen in it to oxidize all organic stuff. Although incineration is thought of as a waste disposal technique, because the ash produced must be disposed of in supervised landfills, it is really a waste reduction technique.

Composting-The process of incinerating garbage involves feeding the screened material through a chamber that is heated to a high temperature (around 1200 °C) and has enough oxygen in it to oxidize all organic stuff. Although incineration is thought of as a waste disposal technique, because the ash produced must be disposed of in supervised landfills, it is really a waste reduction technique.

Water management is the application of techniques, such rainwater gathering and effective irrigation, to adapt to shifting water supply. The design of infrastructure to survive the effects of climate change, such as the construction of sea walls to fend against sea level rise, is known as infrastructure resilience. Community engagement is the process of involving local communities in decision-making and enhancing their ability to adapt through training and education.

10 Future Perspective

Globally, land use and land cover (LUCC) have been dramatically changed by human activity, with severe environmental and socioeconomic repercussions. Sustainable land management requires an understanding of present patterns as well as the ability to forecast future scenarios. Based on existing trends and anticipated future events, the following prognosis is provided:

Urbanization: As urban areas grow quickly, natural habitats and agricultural land are being converted. It is anticipated that this tendency will continue to have an effect on food security, biodiversity, and ecosystem services, particularly in emerging nations. Agriculture Intensification: In order to meet the rising demand for food, agriculture is probably going to becoming more intensive, requiring more water, pesticides, and fertilizers. If sustainable practices are not implemented, this could worsen the state of the soils, water, and biodiversity.

Climate Change: Decisions about land use will be influenced by the effects of climate change, which include extreme weather, rising sea levels, and changes in temperature and precipitation patterns (Bununu et al., 2023) In order to lessen adverse effects on ecosystems, water resources, and agriculture, adaptation measures will be crucial.

Land Degradation: There are still problems with soil erosion, desertification, deforestation, and wetlands degradation. Reforestation, soil conservation, and the restoration of degraded lands are examples of sustainable land management techniques that will be essential in the fight against these trends.

Infrastructure Development: The building of roads, dams, and other infrastructure projects will keep ecosystems impacted, habitats fragmented, and landscapes altered. To reduce adverse effects, development and conservation measures must be balanced.

Policy and Governance: To handle conflicting land use demands, effective land use planning will need to be backed by robust policies and governance. For sustainable land use practices, integrated approaches that take into account social, environmental, and economic considerations will be essential. Technological Innovations: Tools to monitor changes in land use and promote sustainable land management techniques are made possible by technological advancements like precision agriculture, GIS, and remote sensing. Future land use planning will depend on how these technologies are incorporated into the decision-making process.

11 Conclusion

The current study shows that human intervention has a major impact on the changes in land use and cover (LUCC) in Thane, Maharashtra India. A significant decline in agricultural land and wooded regions through field research and remote sensing analysis, mostly as a result of urbanization and industrialization was reported. Numerous environmental issues have resulted from this transition, such as a decline in biodiversity, a rise in greenhouse gas emissions, and modifications to hydrological cycles. The results highlight how urgently the area needs sustainable development plans and efficient land use planning. Reducing the adverse impacts of LUCC can be accomplished by putting policies in place such afforestation, sustainable farming methods, and stringent land use laws. To encourage a sense of environmental responsibility among the local populace, community participation and awareness initiatives are also crucial.

For the purpose of developing focused interventions and conservation initiatives, our study offers insightful information to environmentalists, urban planners, and lawmakers. A multi-stakeholder approach is necessary to address LUCC challenges, with a focus on combining environmental sustainability with economic development.

References

Anand, B., Mariyappan, S., & Rekha, R. S. (2024). Long-term shoreline and LULC change computational analysis in part of the east coast of Tamil Nadu using geoinformation tools. *Journal of Sedimentary Environments, 9*, 707–726. https://doi.org/10.1007/s43217-024-00191-9

Arumugam, N., & Kumaresan, V. (2014). *Environmental Science & Engineering* (2nd ed.). Saras Publication. ISBN 978 9382459590.

Bununu, Y. A., Bello, A., & Ahmed, A. (2023). Land cover, land use, climate change and food security. *Sustainable Earth Reviews, 6*, 16. https://doi.org/10.1186/s42055-023-00065-4

Eslamian, S., Huda, M. B., Rather, N. A., & Eslamian, F. A. (Eds.). (2024). Handbook of climate change impacts on river basin management: Fundamentals and impacts (1st ed.). CRC Press. https://doi.org/10.1201/9781003473336

Evenson, R. E., & Gollin, D. (2003). Assessing the impact of the green revolution 1960 to 2000. *Science*, 300(5620), 758–762.

Foley, J. A., DeFries, R., Asner, G. P., Barford, C., Bonan, G., Carpenter, S. R., Chapin, F. S., Coe, M. T., Daily, G. C., Gibbs, H. K., Helkowski, J. H., Holloway, T., Howard, E. A., Kucharik, C. J., Monfreda, C., Patz, J. A., Prentice, I. C., Ramankutty, N., Snyder, P. K. (2005). Global consequences of land use. *Science, 309*(5734), 570–557.

Geist, H. J., & Lambin, E. F. (2002). Proximate causes and underlying driving forces of tropical deforestation. *BioScience, 52*(2), 143–150.

Khan, M. A., Rahman, A. U., Khan, Z., Sultan, Z., Marwat, F., Naz, T., & Zahid, B. (2024). Evaluating and predicting the land use land cover changes and its impact on land surface temperature using CA-Markov model: A study of District Mardan, Pakistan. *IJIST*, 352–372.

Lambin, E. F., & Geist, H. (2006). *Land-use and land-cover change: Local processes and global impacts*. Springer.

Liu, Z., He, C., & Wu, J. (2016). The relationship between habitat loss and fragmentation during urbanization: An empirical evaluation from 16 world cities. *PLoS ONE, 11*(4), Article e0154613. https://doi.org/10.1371/journal.pone.0154613

Meadows, D. H., Meadows, D. L., & Randers, J. (1992). *Beyond the limits: Global collapse or a sustainable future*. Earthscan.

Muche, M., Yemata, G., Molla, E., Adnew, W., & Muasya, A. M. (2023). Land use and land cover changes and their impact on ecosystem service values in the north-eastern highlands of Ethiopia. *PLoS One, 18*(9), e0289962. https://doi.org/10.1371/journal.pone.0289962. PMID: 37682896; PMCID: PMC10490966.

Nuissl, H., & Siedentop, S. (2021). Urbanisation and land use change. In T. Weith, T. Barkmann, N. Gaasch, S. Rogga, C. Strauß, & J. Zscheischler (Eds.), *Sustainable land management in a European Context*. Human-environment interactions, vol. 8. Springer, Cham. https://doi.org/10.1007/978-3-030-50841-8_5

Ponting, C. (2007). *A new green history of the world: The environment and the collapse of great civilizations*. Penguin Books.

Prem, R., Sheikh, M. M. (2023). Study and analysis of land use/land cover changes of jodhpur city and its impacts on economy and environment (1990–2022). *Journal of Global Resources, 9*(01), 86–95. https://doi.org/10.46587/jgr.2023.v09i01.009

Rai, M. K., Paudel, B., Zhang, Y., Nepal, P., Khanal, N. R., Liu, L., & Rai, R. (2023). Appraisal of empirical studies on land-use and land-cover changes and their impact on ecosystem services in Nepal Himalaya. *Sustainability, 15*(9), 7134. https://doi.org/10.3390/su15097134

Reis, S. (2008). Analyzing land use/land cover changes using remote sensing and GIS in Rize. *North-East Turkey Sensors, 8*(10), 6188–6202.

Shiva, V. (1991). *The violence of the green revolution: Third world agriculture, ecology, and politics*. Zed Books.

Turner, B. L., Lambin, E. F., & Reenberg, A. (2007). The emergence of land change science for global environmental change and sustainability. *Proceedings of the National Academy of Sciences of the United States of America, 104*(52), 20666–20671. http://www.jstor.org/stable/25450958

Wang, Q., Xiong, K., Zhou, J., Xiao, H., & Song, S. (2023). Impact of land use and land cover change on the landscape pattern and service value of the village ecosystem in the karst desertification control. *Frontiers in Environmental Science, 11*, 1020331. https://doi.org/10.3389/fenvs.2023.1020331. ISSN=2296-665X.

Biochar and Sustainable Soil Health

Mahavir Gosavi, Lal Sahab Yadav, and Ashish Jain

Abstract

Biochar is a high carbon-containing product obtained from the thermal decomposition or pyrolysis of organic biomass. Biochar is known for its significance in soil amendment by improving soil fertility, water retention, nutrient cycling, and carbon sequestration. It may modify symbiotic relationships between plants and microbes near the root, thereby enhancing microbial activity and overall soil health. Additionally, biochar's long-lasting carbon sequestration potential contributes to climate change mitigation efforts. In agricultural settings, biochar can significantly increase crop yields by improving nutrient availability and soil fertility. The use of biochar can reduce the application for synthetic fertilizers and pesticides, leading to more sustainable agricultural practices. Biochar's impact on soil health is not limited to conventional agriculture. Due to some important characteristics like porosity, uneven expanded surfaces and chemical makeup, biochar considered as an efficient soil amendment substance. The functions of biochar include improvements in Physical and chemical characteristic of soil. It facilitates macro-aggregation of soil, thus reducing soil erosion and enhancing soil stability. Soil carbon sequestration is also achieved via the carbon content of biochar. While biochar has shown immense potential for enhancing soil health and sustainability, its effects can vary depending on soil type, production method, and application rate. However, further investigation is required to understand these variations and optimize biochar application for specific soil conditions and farming practices. Future studies should focus on refining biochar production methods, understanding its effects on different soil types, and integrating it into diverse agricultural systems to maximize its potential for sustainable soil management.

Keywords

Biochar · Biomass · Pyrolysis · Plant growth · Carbon sequestration · Soil amendment

1 Introduction

Soil health is an important aspect for maintaining agricultural sustainability and ecosystem stability. Healthy soil supports plant growth, stores carbon, filters water, and provides habitat for countless organisms. Unfortunately, our soil faces significant challenges. Erosion, caused by wind, water, and human activities, removes precious top soil. Nutrient depletion occurs when crops remove nutrients faster than they are replaced. Chemical overuse in agriculture can harm soil microbes and lead to salinization, where an excess of salts accumulates in the soil. Climate change further exacerbates these issues, with extreme weather events like droughts and floods impacting soil health (Reicosky, 2020).

Recently, biochar has become recognized as a promising method for sustainably improving soil health. Biochar, a form of charcoal produced from biomass, offers distinctive qualities that enhance soil structure, promote nutrient retention, and foster microbial activity (Kumar & Bhattacharya, 2021).

This chapter aims to explore the major significances and potential benefits of biochar in order to improve agriculture and environmental management practices.

M. Gosavi
Department of Botany, SIES College of Arts, Science & Commerce Sion (W), Mumbai, India

L. S. Yadav
Department of Botany, Smt. CHM College, Ulhasnagar, Maharashtra, India

A. Jain (✉)
Department of Microbiology, Smt. CHM College, Ulhasnagar, Maharashtra, India
e-mail: ajcdri@gmail.com

J. A. Parray (ed.), *Soil and Land Use Change*, Sustainable Landscape Planning and Natural Resources Management,
https://doi.org/10.1007/978-3-031-93075-1_3

2 Biochar: A Brief Overview

- **Definition and Composition**: Biochar is a highly carbon-rich substance produced by thermal decomposition or pyrolysis of organic biomass. Pyrolysis is a process in which heating of organic substances (biomass) in absence of oxygen (Tripathi et al., 2016). It is distinguished by its porous structure and large surface area, qualities that render it a highly effective soil amendment. Biochar typically consists of carbon (ranging from 50 to 90%), along with varying amounts of ash, volatile organic compounds, and other minerals (Bolan et al., 2022). The composition of biochar influenced by type of organic substances as well as specific pyrolysis conditions employed during production.
- **Historical Background and Traditional Uses**: The origins of biochar production and usage trace back to ancient agricultural techniques (Wiedner & Glaser, 2015). Historical evidence suggests that civilizations such as the Amazonian indigenous peoples and ancient Greeks and Romans utilized charcoal-like materials to improve soil fertility and agricultural productivity (Vogel, 2022). For instance, Amazonian dark earth, or terra pieta, is fertile soil enriched with biochar that has sustained agricultural productivity for thousands of years.
- **Contemporary Applications**: In contemporary times, the use of biochar has gain interest due to its potential to tackle diverse agricultural and environmental challenges (Kamali et al., 2022). Some of the contemporary applications of biochar include the following:

  Soil Amendment: Biochar is employed to improve soil fertility, structure, and water retention capacity. Its porous characteristics promote nutrient cycling and provide a home for helpful microbes, which improves agricultural yields and plant health (Alkharabsheh et al., 2021).

  Carbon Sequestration: Biochar, a long-lasting carbon sink, which helps in carbon sequestration and lowering greenhouse gas emissions. It is also stabilizing organic carbon and decreasing the breakdown of organic matter. Biochar contributes to mitigating climate change (Purakayastha et al., 2021).

  Waste Management: Biochar production provides a sustainable approach to managing organic waste streams, which include agricultural residues, forestry residues, and municipal solid waste. By converting biomass into biochar, valuable resources can be recovered while reducing waste disposal and associated environmental impacts (Gupta et al., 2022).

  Environmental Remediation: Biochar has the potential to remediate contaminated soils and water bodies by adsorbing pollutants, thereby reducing their mobility and bioavailability. It has been applied in remediation efforts for heavy metals, organic pollutants, and nutrient runoff in various environmental settings (Das et al., 2023).

3 Biochar: Production Methods

Production methods for biochar vary depending on the desired properties and intended applications. Common production methods include pyrolysis (slow and fast) and gasification.

1. **Slow pyrolysis**: Slow pyrolysis is a thermal process which involves heating biomass to moderate temperatures (often between 300 and 700 °C) without oxygen or in limited supply of oxygen (Nachenius et al., 2013). In this process, organic matter of the biomass is transformed to solid, liquid, and gaseous products via a sequence of chemical reactions including decomposition, vaporization, and carbonization. Slow pyrolysis involves slowly heating biomass at. a controlled rate, due to which volatile compounds release gradually and high-quality biochar formed at the end. The gradual heating and oxygen-free conditions encourage biomass conversion into biochar with stable carbon content and minimal thermal degradation. Slow pyrolysis is preferred for biochar production because it preserves organic carbon in the biochar, creating a stable form of carbon which can remain in soil for extended periods. The resulting biochar is known for its porous structure, large surface area, and advantageous qualities for soil enhancement and environmental uses (Lehmann et al., 2015). Therefore, slow pyrolysis is a crucial technique for generating top-quality biochar with favorable attributes that enhance soil health, promote carbon sequestration, and support environmental sustainability.
2. **Fast pyrolysis**: In fast pyrolysis, quick heating of organic substances at elevated temperatures (400–600 °C) without oxygen or with restricted oxygen resulting fast transformation of biomass into biofuels and biochar (Rasaq et al., 2021). Fast pyrolysis breaks down biomass more quickly by using higher heating rates and shorter residence times than slow pyrolysis. During fast pyrolysis, biomass is rapidly heated to temperatures surpassing its decomposition point, causing rapid thermal cracking and vaporization. This process results in the decomposition of biomass into a combination of gases, liquids, and solids, yielding bio-oils, syngas, and biochar (Uddin et al., 2018). The volatile components released during pyrolysis are condensed into bio-oils, which can be further processed into liquid biofuels or other value-added products. The solid left over residue after pyrolysis process is biochar, which possess significant amount of

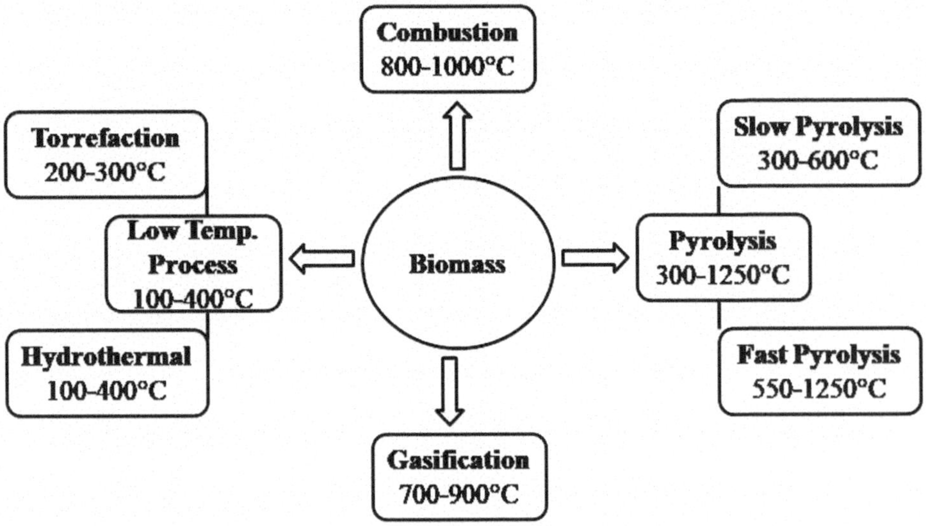

Fig. 1 Schematic representation of different methods to produce biochar

the carbon. Fast pyrolysis offers several advantages over slow pyrolysis, including higher bio-oil yields, shorter processing times, and greater versatility in feedstock options. However, the biochar produced via fast pyrolysis may have different properties compared to slow pyrolysis biochar, including lower surface area and higher ash content. However, fast pyrolysis continues to be a critical technology for producing biofuels and generating biochar, thereby contributing to production of renewable energy, carbon sequestration and sustainable waste management (Makepa et al., 2023).

Gasification: It is a thermo-chemical process that transforms organic substances into syngas (synthesis gas) through partial combustion at high temperatures (700–1200 °C) in the presence of limited oxygen (Pandey et al., 2022). In the process of gasification, biomass undergoes various chemical reactions, including pyrolysis, oxidation, and reduction, leading to the transformation of organic matter into gaseous form which contains mixture of H, CO_2, CO, & CH_4 (Akhtar et al., 2018). The composition of the syngas depend on the feedstock composition, gasification circumstances, and other process parameters. Syngas derived from biomass gasification serves multiple purposes, including heat and power generation, fuel production, and chemical synthesis. Moreover, gasification can integrate with biochar production, where the leftover solid residue from syngas production can be further refined into biochar (Yao et al., 2018). The steps involve in the process of biochar production are.

- **Biomass Feedstock Preparation**: The biomass feedstock, such as agricultural wastes, agro-industrial wastes, and other organic wastes, undergoes preparation and drying to decrease moisture content (Wiselogel et al., 2018).
- **Gasification**: The prepared biomass is introduced into a gasifier, where it undergoes partial combustion at elevated temperatures in presence of required amount of oxygen and/or heat steam. The gasification process leads to the creation of syngas, which exits the gasifier as a gaseous mixture (Kan & Strezov, 2014).
- **Syngas Cleaning**: The crude syngas produced from gasification contain impurities such as tar, particulate matter, and sulfur compounds. Syngas cleaning processes, such as catalytic conversion, filtration and scrubbing are employed to removal of contaminants and enhancement the purity and characteristics of the syngas (de Oliveira et al., 2023).

Biochar Production: The solid residue, referred to as char or biomass ash, left after syngas production can undergo additional processing into biochar using pyrolysis or carbonization techniques. This process involves heating the char at high temperatures without oxygen in order to convert it into a stable, carbon-rich material that is ideal for environmental applications and soil enhancement (Ghesti et al., 2022). (Fig. 1)

4 Classification

Biochar can be categorized into different types based on various criteria such as raw material used, production method, and intended application (Seow et al., 2022). Here are some common types of biochar:

1. **Feedstock-based classification**:

   - **Woody biochar**: Made from wood and woody biomass such as branches, sawdust, and wood chips.

- **Agricultural waste biochar**: Produced from agricultural residues like straw, husks, stalks, and other crop residues.
- **Manure-based biochar**: Derived from animal manure, such as poultry litter or cow dung.
- **Algae-based biochar**: Made from algae biomass, which can be derived from wastewater treatment plants or natural aquatic systems.

2. **Production method-based classification**:

- **Pyrolysis biochar**: It is produced by slow and fast pyrolysis process by heating biomass in the absence or limited oxygen, resulting in the production of biochar as by product.
- **Hydrothermal carbonization (HTC) biochar**: It is produced by the process of hydrothermal carbonization, which includes heating biomass at moderate temperatures and pressures in the presence of water.
- **Gasification biochar**: Generated through the gasification process, which involves partial oxidation of biomass to produce syngas, with biochar as a byproduct.
- **Microwave-assisted biochar**: Produced using microwave heating technology, which offers rapid and efficient pyrolysis of biomass.

3. **Application-based classification**:

- **Soil amendment biochar**: Specially engineered to enhance soil properties, such as improving water retention, increasing nutrient availability, and enhancing soil structure.
- **Water treatment biochar**: Utilized for water purification by adsorption of pollutants such as heavy metals and organic pollutants from water.
- **Energy biochar**: Primarily produced for energy generation, often as a co-product of bioenergy production processes like pyrolysis and gasification.
- **Livestock feed additive biochar**: Utilized as a feed supplement for livestock to improve digestion, reduce odor, and enhance nutrient utilization.
- **Environmental remediation biochar**: Applied in remediation projects to mitigate soil and water pollution, such as in contaminated land restoration and mine site reclamation.

5 Biochar Structural Properties

Biochar's distinctive structural properties contribute to its advantageous effects in soil improvement, carbon sequestration, and environmental remediation. These properties include:

- **Porosity**: Biochar exhibits significant porosity characterized by a sophisticated network of pores at micro-, meso-, and macroscopic scales. This intricate pore structure offers an expansive surface area, facilitating the absorption of water, nutrients, and contaminants. The porosity of biochar contributes to facilitate improved aeration, water retention, and increased nutrient accessibility of soil, ultimately fostering plant growth and enhancing soil vitality (Hue, 2020).
- **Surface Area**: The uneven surface area and porous nature of biochar enhances ability to adsorb and retain molecules, ions, and gases. The surface area significantly impacts its reactivity, capacity for sorption, and effectiveness across different applications (Leng et al., 2021).
- **Surface Chemistry**: The chemical nature of biochar surface is shaped by the feedstock used, the conditions of pyrolysis, and by other subsequent processing treatments. The surface is highly reactive due to carboxyl (–COOH), hydroxyl (–OH) and phenolic (–Ph) functional group, which participates in chemical reactions, ion exchange, and sorption processes. These functional groups influence the cation exchange capacity (CEC), pH and sorption affinity of biochar (Thomas, 2021).
- **Stability**: Biochar represents a durable and resilient carbon form that shows remarkable resistance to decomposition and breakdown in soil as time progresses. This enduring stability is linked to several factors including carbonization temperature, mineral content, and chemical makeup. Due to its stability, biochar can endure in soil for many years, functioning as a persistent carbon sink and beneficial soil amendment (Purakayastha et al., 2015).
- **Particle size distribution**: The size and shape of biochar particles vary significantly, due to nature of feedstock, pyrolysis condition and some subsequent post-processing techniques. The distribution of particle sizes directly impacts the physical characteristics of biochar, including its bulk density, capacity to retain water, and suitability for soil integration. Fine-grained biochar typically offers a larger surface area and increased reactivity, while coarse-grained biochar tends to enhance soil structure and aeration (Ramanayaka et al., 2020).

6 Biochar Nutrients

1. Biochar, characterized by its stable carbon structure, generally does not directly supply nutrients to plants. However, the nutrient content of biochar influenced by biomass and the temperature employed during its production. Biochar obtained from manure or biosolids often contains high amount of nitrogen, phosphorus and potassium compared to biochar produced from woody materials (Joseph et al., 2021).
2. Biochar's notable strength lies in its capacity to retain nutrients within the soil. Its porous structure functions akin to a miniature sponge, effectively holding onto essential nutrients and reduces leaching (Dwibedi et al., 2022).
3. Biochar can also indirectly improve nutrient availability for plants by fostering a healthy soil environment. The porous surface structure provides suitable habitat for soil microbes, playing a key role in cycling of nutrients. These microbes decompose organic matter and transform nutrients into forms that are more accessible for plant uptake (Gorovtsov et al., 2020).

7 Biochar's Impact on Soil Health

Physical Properties and Effects on Soil Structure: The physical characteristics of soil are significantly impacted by biochar due to porous structure and expansive surface area (Zhang et al., 2021). These properties influence soil structure in several ways:

- **Improved Soil Porosity**: Biochar's porous structure enhances soil porosity, increasing air and water movement within the soil. This improves soil aeration, root penetration, and water infiltration, particularly in compacted or poorly draining soils.
- **Enhanced Water Retention**: The surface of biochar possesses numerous pores which make it suitable to absorb and retain water, reduces water runoff and enhances water availability for plants. This characteristic is especially advantageous in sandy soils that have a low water-holding capacity.
- **Reduced Soil Erosion**: Biochar can aid in preventing soil erosion by stabilizing soil aggregates and decreasing surface runoff. Its presence in the soil can enhance soil stability and reduce the risk of erosion caused by wind or water.

1. **Chemical Properties and Impacts on Nutrient Cycling**: Biochar influences soil nutrient dynamics through its chemical properties, which include high cation exchange capacity (CEC), pH buffering capacity, and nutrient retention capabilities (Antonangelo et al., 2024). These properties affect nutrient cycling in the following ways:

    - **Nutrient Retention**: Biochar exhibits a strong affinity for macro and micronutrients. It adsorb these nutrients and gradually release them over time, thereby increasing their availability to plants and minimizing nutrient leaching.
    - **pH Buffering**: Biochar can buffer soil pH, assisting in the maintenance of optimal pH levels for plant growth. In acidic soils, biochar can raise pH levels, while in alkaline soils, it can lower pH levels, thereby enhancing nutrient availability and subsequently uptake by plants.
    - **Cation Exchange Capacity (CEC)**: The surface of biochar possesses several functional groups which enables it to adsorb and retain cations like calcium (Ca), magnesium (Mg), and potassium (K). This ability helps stabilize soil nutrients and mitigate their loss through leaching.

2. **Biological Properties and Effects on Soil Microbial Activity**:

    Biochar influences soil microbial communities and activity through various mechanisms, including its physical structure, chemical composition, and interactions with microorganisms (Singh et al., 2022).

8 Biochar and Soil Carbon Sequestration

1. **Mechanisms of Carbon Sequestration**: Biochar contributes to soil carbon sequestration through several mechanisms (Lal, 2016):

    - **Stable Carbon Storage**: Biochar is an extremely stable carbon form that can stay in the soil for hundreds to thousands of years without decomposing. By burying carbon in the soil, biochar effectively removes atmospheric carbon dioxide (CO_2) from the carbon cycle, mitigating greenhouse gas emissions.

- **Carbon Saturation**: The substantial surface area and porous nature of biochar offer abundant sites for carbon adsorption and stabilization. As biochar ages in the soil, it continues to accumulate carbon through interactions with soil minerals, organic matter, and microbial residues, gradually increasing soil carbon stocks over time.
- **Reduced Soil Carbon Losses**: Biochar can assist in reducing soil carbon losses by improving soil structure, facilitating microbial stabilization of organic matter, and decreasing the breakdown of soil organic carbon. These effects lead to enhanced soil carbon retention and stability, thereby contributing to carbon sequestration efforts.

2. **Effect of Biochar on Climate Change**: Biochar holds promising potential for mitigating climate change through carbon sequestration in the soil and the reduction of greenhouse gas emissions (Fawzy et al., 2021):

   - **Carbon Offset**: The carbon stored in biochar represents a net removal of carbon dioxide from the atmosphere, effectively offsetting emissions from fossil fuel combustion and other sources. By enhancing soil carbon stocks, biochar can contribute to achieving carbon neutrality and meeting emissions reduction targets.
   - **Climate Resilience**: Elevated soil carbon levels due to biochar application can improve soil fertility, enhance water retention, and increase crop productivity. These benefits can help in mitigate of climate change.
   - **Co-benefits for Sustainable Development**: Biochar production and application offer additional co-benefits beyond carbon sequestration, including waste management, soil improvement, and renewable energy production. These co-benefits can support sustainable development goals while mitigating climate change.

3. **Challenges and Limitations of Biochar as a Carbon Sink**: Despite its potential benefits, biochar faces several challenges and limitations as a carbon sequestration tool (Elkhlifi et al., 2023):

   - **Feedstock Availability and Sustainability**: Securing adequate biomass feedstocks for biochar production, especially at larger scales, may pose challenges due to limited availability. It's crucial to ensure sustainable sourcing of feedstocks to prevent competition with food production, avoid deforestation, and mitigate other environmental impacts.
   - **Economic Viability**: The economic viability, production and application of biochar is influenced by factors such as feedstock costs, processing technology, and market demand. High initial investment costs and uncertain market incentives could impede the extensive adoption of biochar as a carbon sequestration solution.
   - **Environmental Risks and Uncertainties**: While biochar can enhance soil fertility and productivity, its long-term environmental impacts are still being studied. Potential risks include soil acidification, nutrient imbalances, and changes in soil microbial communities. Assessing and mitigating these risks is critical to ensure the sustainability of biochar applications.
   - **Policy and Regulatory Challenges**: The development of supportive policies and regulatory frameworks is essential to incentivize biochar adoption and ensure environmental sustainability. Clear guidelines for biochar production, application, and monitoring can help address potential environmental and social concerns.

Effects of Biochar on Plant Growth and Yield: Application of Biochar can have positive effects on plant growth, productivity, and overall crop performance. These effects are attributed to various mechanisms, including physical and chemical nature of soil (Dai et al., 2020).

- **Enhanced Root Development**: Biochar's porous structure and enhanced soil aeration promote root growth and proliferation, allowing plants to access nutrients and water more effectively.
- **Increased Nutrient Uptake**: Biochar in the soil improves nutrient availability and uptake by plants, resulting in enhanced nutrient status and growth.
- **Water Retention**: Biochar's water-holding capacity helps to maintain soil moisture levels and reduce water stress on plants, particularly during periods of drought or water scarcity.
- **Disease Suppression**: Biochar can suppress soil-borne pathogens and diseases by altering soil microbial communities and enhancing plant defense mechanisms, thereby improving plant health and resilience.
- **Yield Improvement**: In general, the application of biochar has demonstrated an increase in crop yields and improved crop quality across various agricultural systems, including arable crops, horticulture, and forestry.

Implications for Drought Resilience and Water Management: Biochar's water retention properties have significant implications for drought resilience and water management in agricultural systems (Ndede et al., 2022):

- **Drought Resilience**: Biochar-amended soils are more resilient to drought stress due to their enhanced water-holding capacity. Plants grown in biochar-amended soils are better able to withstand periods of water scarcity, as biochar assists in maintaining optimal soil moisture levels essential for plant growth and development.
- **Water Conservation**: Biochar application can contribute to water conservation efforts by reducing irrigation requirements and minimizing water loss through evaporation and runoff. This can result in more efficient water use and lower water consumption in agricultural production systems.
- **Improved Crop Performance**: By improving soil moisture availability, biochar can enhance crop performance and productivity, even under water-limited conditions. Plants grown in biochar-amended soils may exhibit increased biomass accumulation, yield stability, and tolerance to water stress.

9 Biochar and Soil Contaminant Remediation

1. **Adsorption of Contaminants**: The substantial surface area and porosity making it an effective adsorbent for application. The adsorption process involves the physical or chemical attraction of contaminants to biochar surfaces, where they are immobilized and held in place (Qiu et al., 2022).

   - **Organic Contaminants**: Biochar has the ability to adsorb a diverse array of organic contaminants, such as pesticides, herbicides, petroleum hydrocarbons, and organic solvents.The aromatic carbon structures and functional groups present on biochar surfaces facilitate the adsorption of organic molecules through hydrophobic and π-π interactions.
   - **Inorganic Contaminants**: Biochar is also capable of adsorbing inorganic contaminants such as heavy metals, metalloids, and ions. The negatively charged surfaces of biochar particles attract positively charged metal ions through electrostatic interactions, reducing their mobility and bioavailability in the soil.

2. **Effects on Soil pH and Redox Potential**: The application of biochar can impact soil pH and redox potential, which subsequently influence the mobility and transformation of contaminants within the soil environment (Mandal et al., 2021).

   - **pH Buffering**: Biochar has pH buffering capacity, meaning it can help stabilize soil pH and prevent drastic fluctuations. In acidic soils, biochar can raise pH, while in alkaline soils, it can lower pH, bringing it closer to neutral. The pH buffering effect of biochar can influence the speciation and mobility of contaminants in the soil.
   - **Redox Reactions**: Biochar can influence soil redox conditions by altering oxygen availability and microbial activity. In anaerobic environments, biochar an serve as an electron acceptor, promoting reducing conditions and facilitating the immobilization of certain contaminants, such as heavy metals.

3. **Applications in Contaminated Land Remediation**: Biochar has shown promising applications in the remediation of contaminated lands and environments, offering a sustainable and cost-effective approach to pollutant removal and soil restoration (Haider et al., 2022).

   - **In Situ Stabilization**: Biochar can be directly applied to contaminated soils as a soil amendment to immobilize and sequester contaminants in place. By adsorbing contaminants and reducing their bioavailability, biochar helps to mitigate the risks of human exposure and environmental contamination.
   - **Phytoremediation Enhancement**: Biochar can enhance phytoremediation processes by improving soil structure, nutrient availability, and microbial activity. When combined with phytoremediation techniques, biochar can facilitate the uptake and accumulation of contaminants by plants, leading to their removal from the soil through plant uptake and biomass harvest.
   - **Water Treatment**: Biochar can be utilized in water treatment systems to effectively remove contaminants from polluted water sources, such as industrial wastewater, agricultural runoff, and mine drainage. By adsorbing pollutants and improving water quality, biochar helps to protect aquatic ecosystems and human health.

10 Biochar and Sustainable Agriculture Practices

1. **Integrating Biochar into Agricultural Systems**: Biochar can be integrated into agricultural systems through various application methods and practices (Schmidt et al., 2021):

   - **Soil Amendment**: The most common application of biochar in agriculture involves incorporating it into the soil as a soil amendment. Biochar can be mixed into the soil during land preparation, applied as a top dressing, or incorporated into planting holes during crop establishment.
   - **Composting**: Biochar can be added to composting systems to enhance compost quality and stability. Mixing biochar with organic waste materials during composting helps to improve nutrient retention, microbial activity, and carbon stabilization, resulting in higher-quality compost for soil amendment.
   - **Mulching**: Biochar can be applied as a mulch material to improve nutrient availability, inhibit weed growth, and preserve soil moisture. This will eventually encourage plant growth and reduce soil erosion

2. **Best Practices and Guidelines for Biochar Application**: In order to maximize beneficial effects of biochar in agriculture, it is essential to follow best practices and guidelines for its application (Mabrouk et al., 2023):

   - **Feedstock Selection**: Choose high-quality feedstocks for biochar production, such as woody biomass, crop residues, or organic wastes. Avoid feedstocks contaminated with pollutants or treated with harmful chemicals.
   - **Pyrolysis Conditions**: To produce biochar with desirable properties, it is important to optimize pyrolysis conditions, which include temperature, heating rate, and residence time. Slow pyrolysis at moderate temperatures (typically between 400 and 600 °C) often results in high-quality biochar with stable carbon content and beneficial properties.
   - **Particle Size and Distribution**: Use biochar particles of appropriate size and distribution to ensure uniform incorporation into the soil. Fine-grained biochar may provide greater surface area and reactivity, while coarse-grained biochar may improve soil structure and aeration.
   - **Application Rates**: Determine appropriate application rates based on soil type, crop requirements, and desired outcomes. Start with conservative application rates and monitor soil response before adjusting application rates accordingly.
   - **Incorporation Methods**: Ensure thorough mixing and incorporation of biochar into the soil to promote uniform distribution and maximize contact with soil particles. Incorporation methods may include mechanical mixing, tillage, or irrigation-based application.

3. **Case Studies and Success Stories in Biochar-Enhanced Agriculture**: There is a wealth of case studies and success stories showcasing the benefits of biochar in improving agricultural productivity, enhancing soil health, and promoting environmental sustainability:

   - **Terra Preta**: The ancient Amazonian practice of adding biochar to soils, known as terra preta, has sustained fertile and productive soils in the region for thousands of years. Studies have shown that terra preta soils exhibit higher fertility, carbon content, and crop yields compared to adjacent soils (Lehmann, 2009).
   - **Biochar Trials**: Field trials and research projects conducted worldwide have consistently demonstrated the positive impacts of biochar on crop yields, nutrient cycling, and soil health. From smallholder farms to large-scale commercial agriculture operations, biochar has been shown to enhance soil fertility, improve water retention, and boost crop performance across diverse agroecosystems (Diatta et al., 2020).
   - **Sustainable Land Management**: Biochar has been effectively incorporated into sustainable land management practices, including agroforestry, organic farming, and conservation agriculture. By enhancing soil fertility and resilience, biochar plays a crucial role in promoting the long-term sustainability and productivity of agricultural landscapes (Blanco-Canqui, 2021).

11 Conclusion

In conclusion, biochar represents a promising and sustainable solution for enhancing soil health, agricultural productivity, and environmental resilience. Throughout this paper, we have explored the multifaceted role of biochar in soil health improvement, carbon sequestration, water retention, contaminant remediation, and microbial activity promotion.

Key points highlighted include the following:

- Biochar's impact on soil physical, chemical, and biological properties encompasses its effects on soil structure, nutrient cycling, pH buffering, and microbial activity.

- Biochar plays an important role in sequestration of carbon, mitigating climate change and reducing greenhouse gas emissions.
- Biochar's effectiveness in retaining water in soil, improving soil moisture dynamics, and enhancing drought resilience in agricultural systems.
- Its capacity to adsorb contaminants, influence soil redox potential and support remediation efforts in contaminated lands and environments.
- The promotion of beneficial soil microbes by biochar, leading to improved plant health, disease suppression, and nutrient cycling in agricultural soils.

12 Prospects and Research Directions

Looking ahead, prospects and research directions for biochar include the following:

- Further exploration of biochar production methods, feedstock selection, and pyrolysis conditions to optimize biochar properties and performance.
- Long-term field studies and ongoing monitoring efforts are essential for assessing the impacts of biochar on environmental, soil health, crop yields, and ecosystem services.
- The integration of biochar into sustainable land management practices, agroecosystem design, and climate-smart agriculture initiatives is key to maximizing its potential benefits while minimizing potential risks.
- Development of guidelines, policies, and incentives to promote responsible biochar production, application, and utilization in agricultural and environmental contexts.

Overall, biochar holds great potential as a sustainable soil health solution, offering wide approach to tackle key challenges in modern agriculture, climate change mitigation, and environmental remediation. By harnessing the benefits of biochar and incorporating it into soil management practices, stakeholders can work toward building resilient and productive agricultural systems while mitigating environmental impacts.

References

Akhtar, A., Krepl, V., & Ivanova, T. (2018). A combined overview of combustion, pyrolysis, and gasification of biomass. *Energy & Fuels, 32*(7), 7294–7318.

Alkharabsheh, H. M., Seleiman, M. F., Battaglia, M. L., Shami, A., Jalal, R. S., Alhammad, B. A., Almutairi, K. F., & Al-Saif, A. M. (2021). Biochar and its broad impacts in soil quality and fertility, nutrient leaching and crop productivity: A review. *Agronomy, 11*(5), 993.

Antonangelo, J. A., Culman, S., & Zhang, H. (2024). Comparative analysis and prediction of cation exchange capacity via summation: Influence of biochar type and nutrient ratios. *Frontiers in Soil Science, 4*, 1371777.

Blanco-Canqui, H. (2021). Does biochar improve all soil ecosystem services? *GCB Bioenergy, 13*(2), 291–304.

Bolan, N., Hoang, S. A., Beiyuan, J., Gupta, S., Hou, D., Karakoti, A., & Joseph, S. (2022). Multifunctional applications of biochar beyond carbon storage. *International Materials Reviews, 67*(2), 150–200.

Dai, Y., Zheng, H., Jiang, Z., & Xing, B. (2020). Combined effects of biochar properties and soil conditions on plant growth: A meta-analysis. *Science of the Total Environment, 713*, Article 136635.

Das, S. K., Ghosh, G. K., & Avasthe, R. (2023). Biochar application for environmental management and toxic pollutant remediation. *Biomass Conversion and Biorefinery, 13*(1), 555–566.

de Oliveira, D. C., Lora, E. E., Venturini, O. J., Maya, D. M., & Garcia-Pérez, M. (2023). Gas cleaning systems for integrating biomass gasification with Fischer-Tropsch synthesis-A review of impurity removal processes and their sequences. *Renewable and Sustainable Energy Reviews, 172*, Article 113047.

Diatta, A. A., Fike, J. H., Battaglia, M. L., Galbraith, J. M., & Baig, M. B. (2020). Effects of biochar on soil fertility and crop productivity in arid regions: A review. *Arabian Journal of Geosciences, 13*, 1–17.

Dwibedi, S. K., Pandey, V. C., Divyasree, D., & Bajpai, O. (2022). Biochar-based land development. *Land Degradation & Development, 33*(8), 1139–1158.

Elkhlifi, Z., Iftikhar, J., Sarraf, M., Ali, B., Saleem, M. H., Ibranshahib, I., Bispo, M. D., Meili, L., Ercisli, S., Torun Kayabasi, E., & Alemzadeh Ansari, N. (2023). Potential role of biochar on capturing soil nutrients, carbon sequestration and managing environmental challenges: A review. *Sustainability, 15*(3), 2527.

Fawzy, S., Osman, A. I., Yang, H., Doran, J., & Rooney, D. W. (2021). Industrial biochar systems for atmospheric carbon removal: A review. *Environmental Chemistry Letters, 19*, 3023–3055.

Ghesti, G. F., Silveira, E. A., Guimarães, M. G., Evaristo, R. B., & Costa, M. (2022). Towards a sustainable waste-to-energy pathway to pequi biomass residues: Biochar, syngas, and biodiesel analysis. *Waste Management, 143*, 144–156.

Gorovtsov, A. V., Minkina, T. M., Mandzhieva, S. S., Perelomov, L. V., Soja, G., Zamulina, I. V., Rajput, V. D., Sushkova, S. N., Mohan, D., & Yao, J. (2020). The mechanisms of biochar interactions with microorganisms in soil. *Environmental Geochemistry and Health, 42*, 2495–2518.

Gupta, M., Savla, N., Pandit, C., Pandit, S., Gupta, P. K., Pant, M., Khilari, S., Kumar, Y., Agarwal, D., Nair, R. R., & Thomas, D. (2022). Use of biomass-derived biochar in wastewater treatment and power production: A promising solution for a sustainable environment. *Science of the Total Environment, 825*, Article 153892.

Haider, F. U., Wang, X., Zulfiqar, U., Farooq, M., Hussain, S., Mehmood, T., Naveed, M., Li, Y., Liqun, C., Saeed, Q., & Ahmad, I. (2022). Biochar application for remediation of organic toxic pollutants in contaminated soils; An update. *Ecotoxicology and Environmental Safety, 248*, Article 114322.

Hue, N. (2020). Biochar for maintaining soil health. *Soil health*, 21–46.

Joseph, S., Cowie, A. L., Van Zwieten, L., Bolan, N., Budai, A., Buss, W., Cayuela, M. L., Graber, E. R., Ippolito, J. A., Kuzyakov, Y., & Luo, Y. (2021). How biochar works, and when it doesn't: A review of mechanisms controlling soil and plant responses to biochar. *Gcb Bioenergy, 13*(11), 1731–1764.

Kamali, M., Sweygers, N., Al-Salem, S., Appels, L., Aminabhavi, T. M., & Dewil, R. (2022). Biochar for soil applications-sustainability aspects, challenges and future prospects. *Chemical Engineering Journal, 428*, Article 131189.

Kan, T., & Strezov, V. (2014). Gasification of biomass. *Biomass processing technologies*, 81–121.

Kumar, A., & Bhattacharya, T. (2021). Biochar: A sustainable solution. *Environment, Development and Sustainability, 23*, 6642–6680.

Lal, R. (2016). Biochar and soil carbon sequestration. *Agricultural and Environmental Applications of Biochar: Advances and Barriers, 63*, 175–197.

Lehmann, J., Abiven, S., Kleber, M., Pan, G., Singh, B. P., Sohi, S. P., & Zimmerman, A. R. (2015). Persistence of biochar in soil. In *Biochar for environmental management* (pp. 235–282). Routledge.

Lehmann, J. (2009). Terra preta Nova–where to from here. *Amazonian dark earths: Wim Sombroek's vision*, 473–486.

Leng, L., Xiong, Q., Yang, L., Li, H., Zhou, Y., Zhang, W., Jiang, S., Li, H., & Huang, H. (2021). An overview on engineering the surface area and porosity of biochar. *Science of the Total Environment, 763*, Article 144204.

Mabrouk, O., Hamdi, H., Sayadi, S., Al-Ghouti, M. A., Abu-Dieyeh, M. H., & Zouari, N. (2023). Reuse of sludge as organic soil amendment: Insights into the current situation and potential challenges. *Sustainability, 15*(8), 6773.

Makepa, D. C., Chihobo, C. H., & Musademba, D. (2023). Advances in sustainable biofuel production from fast pyrolysis of lignocellulosic biomass. *Biofuels, 14*(5), 529–550.

Mandal, S., Pu, S., Adhikari, S., Ma, H., Kim, D. H., Bai, Y., & Hou, D. (2021). Progress and future prospects in biochar composites: Application and reflection in the soil environment. *Critical Reviews in Environmental Science and Technology, 51*(3), 219–271.

Nachenius, R. W., Ronsse, F., Venderbosch, R. H., & Prins, W. (2013). Biomass pyrolysis. In *Advances in chemical engineering* (vol. 42, pp. 75–139). Academic Press.

Ndede, E. O., Kurebito, S., Idowu, O., Tokunari, T., & Jindo, K. (2022). The potential of biochar to enhance the water retention properties of sandy agricultural soils. *Agronomy, 12*(2), 311.

Pandey, P. H., Ukarde, T. M., Mahale, J. S., Vasishta, A., & Pawar, H. S. (2022). Syngas production via biomass gasification. In *Biomass, biofuels, biochemical* (pp. 211–261). Elsevier.

Purakayastha, T. J., Kumari, S., Sasmal, S., & Pathak, H. (2015). Biochar carbon sequestration in soil-A myth or reality? *International Journal of Bio-Resource and Stress Management, 6*(5), 623–630.

Purakayastha, T. J., Bhaduri, D., & Singh, P. (2021). Role of biochar on greenhouse gas emissions and carbon sequestration in soil: Opportunities for mitigating climate change. *Soil Science: Fundamentals to Recent Advances*, 237–260.

Qiu, M., Liu, L., Ling, Q., Cai, Y., Yu, S., Wang, S., Fu, D., Hu, B., & Wang, X. (2022). Biochar for the removal of contaminants from soil and water: A review. *Biochar, 4*(1), 19.

Ramanayaka, S., Vithanage, M., Alessi, D. S., Liu, W. J., Jayasundera, A. C., & Ok, Y. S. (2020). Nanobiochar: Production, properties, and multifunctional applications. *Environmental Science: Nano, 7*(11), 3279–3302.

Rasaq, W. A., Golonka, M., Scholz, M., & Białowiec, A. (2021). Opportunities and challenges of high-pressure fast pyrolysis of biomass: A review. *Energies, 14*(17), 5426.

Reicosky, D. (2020). Conservation agriculture systems: Soil health and landscape management. In *Advances in conservation agriculture* (pp. 87–154). Burleigh Dodds Science Publishing.

Schmidt, H. P., Kammann, C., Hagemann, N., Leifeld, J., Bucheli, T. D., Sánchez Monedero, M. A., & Cayuela, M. L. (2021). Biochar in agriculture–A systematic review of 26 global meta-analyses. *GCB Bioenergy, 13*(11), 1708–1730.

Seow, Y. X., Tan, Y. H., Mubarak, N. M., Kansedo, J., Khalid, M., Ibrahim, M. L., & Ghasemi, M. (2022). A review on biochar production from different biomass wastes by recent carbonization technologies and its sustainable applications. *Journal of Environmental Chemical Engineering, 10*(1), Article 107017.

Singh, H., Northup, B. K., Rice, C. W., & Prasad, P. V. (2022). Biochar applications influence soil physical and chemical properties, microbial diversity, and crop productivity: A meta-analysis. *Biochar, 4*(1), 8.

Thomas, S. C. (2021). Post-processing of biochars to enhance plant growth responses: A review and meta-analysis. *Biochar, 3*(4), 437–455.

Tripathi, M., Sahu, J. N., & Ganesan, P. (2016). Effect of process parameters on production of biochar from biomass waste through pyrolysis: A review. *Renewable and Sustainable Energy Reviews, 55*, 467–481.

Uddin, M. N., Techato, K., Taweekun, J., Rahman, M. M., Rasul, M. G., Mahlia, T. M. I., & Ashrafur, S. M. (2018). An overview of recent developments in biomass pyrolysis technologies. *Energies, 11*(11), 3115.

Vogel, P. (2022). *Sustainable development: The roles of carbon and biocarbon: An introduction to molecular sciences*. pp. 95–131. https://doi.org/10.1142/9789811240492_0003

Wiedner, K., & Glaser, B. (2015). Traditional use of biochar. In *Biochar for environmental management* (pp. 15–37). Routledge.

Wiselogel, A., Tyson, S., & Johnson, D. (2018). Biomass feedstock resources and composition. In *Handbook on bioethanol* (pp. 105–118). Routledge.

Yao, Z., You, S., Ge, T., & Wang, C. H. (2018). Biomass gasification for syngas and biochar co-production: Energy application and economic evaluation. *Applied Energy, 209*, 43–55.

Zhang, Y., Wang, J., & Feng, Y. (2021). The effects of biochar addition on soil physicochemical properties: A review. *CATENA, 202*, 105284.

Sustainable Soil Management—Principles and Importance

T. Dhanalakshmi and J. Rajichellam

Abstract

Soil is a non-renewable resource; however, it takes centuries to form 1 mm of soil. Soil performs many ecosystem functions and services and mainly provides food and supports nutrition. The soil resource has become progressively neglected and threatened in the quest for increased food production, urbanization, and development. Soil health has been vulnerable for many reasons, like soil infertility and degradation, a lack of soil organic carbon and biodiversity, acidification, contamination, salinization, erosion, and degradation of soil. Hence, sustainable soil management is important to protect this valuable natural resource. As per the FAO, soil management can be considered sustainable if it doesn't harmfully disturb biodiversity or the functions of soil that facilitate the abovementioned services. It can also be sustained or improved. This chapter delivers on sustainable soil management principles and their importance in agricultural and environmental services.

Keywords

Soil · Management · Important · Agricultural · Nutrition · Protect

T. Dhanalakshmi (✉)
Nehru Institute of Health Sciences, Coimbatore, Tamil Nadu, India
e-mail: dhanvarshar@gmail.com

J. Rajichellam
Nehru Institute of Engineering and Technology, Coimbatore, Tamil Nadu, India

1 Introduction

The goal of sustainable soil management is to care for the soil in a way that will prolong its health and productivity. This entails using techniques that preserve the structure of the soil, such as minimizing soil disturbance during ploughing and avoiding the use of large equipment that compacts the earth. Adding organic matter, like compost or plant leftovers, to the soil also helps maintain its richness and vibrancy. Plants may acquire the nutrients they require without polluting the environment when the proper amount of fertilizer is applied, as determined by soil testing (Pretty, 2008).

Water management is a critical component of sustainable soil management as well. This involves conserving water by utilizing effective irrigation methods and maintaining soil moisture through mulching (El-Beltagi et al., 2022). Soil retention is aided by growing cover crops and erecting barriers to prevent erosion. Growing a range of plants and safeguarding helpful insects are important ways to promote biodiversity. Together, these methods sustain the health of the soil, encourage plant development, and safeguard the environment (Sharma et al., 2018).

2 Key Principles of Sustainable Soil Management

2.1 Soil components

Understanding the components of soil is essential in order to know the complexity of soil health in numerous agricultural systems. Solid particles are formed by the soil-forming processes called soil. Based on the size of the particles and shape, which can be categorized as clay, silt, or sand. The largest particles are sand (0.05–2 mm), followed by silt (0.002–0.05 mm), and the smallest are clay (<0.002 mm).

J. A. Parray (ed.), *Soil and Land Use Change*, Sustainable Landscape Planning and Natural Resources Management,
https://doi.org/10.1007/978-3-031-93075-1_4

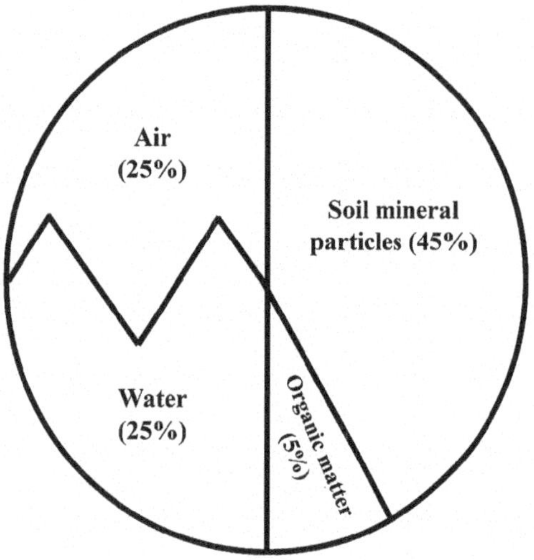

Fig. 1 Soil components

Depending on the condition of the soil, Soils have pore gaps between these particles and that can be filled with water or air (Fig. 1). In disparity to fine-textured or "heavy" soils, which have fewer particles and pores, soils with more sand will have larger pores (Idowu et al., 2020).

Soil texture is based on the relative composition of sand and clay (Fig. 2). Soil has distinct texture, based on the percentage of various-sized particles measured after eliminating all rocks, cobbles, and gravel greater than 2 mm. Textures differ from those of sand, which primarily consists of sand-sized particles, to those of clay, which primarily consists of very minute particles. Roughly 45 percent of the volume of soil is composed of soil particles (Fig. 1) (Idowu et al., 2020).

Fig. 2 Soil textural triangle (USDA Natural Resources Conservation Service)

2.2 Maintaining Soil Structure and Health

Keeping the soil in good condition to support healthy plant growth is known as maintaining soil structure and health. Preventing soil compaction is crucial because it occurs when heavy machinery or excessive foot traffic compacts the soil, making it more difficult for air and water to get through (Tracy et al., 2011). Farmers have access to specialized techniques such as limited tillage, which involves minimizing soil disturbance, and controlled traffic farming, which involves operating their equipment only on designated routes. This keeps the soil loose and full of the tiny openings that roots require in order to develop (Prajapati et al., 2023).

For healthy soil, organic matter must be added, such as compost or leftover plant material. Little organisms in the soil, including worms and bacteria, are fed by this organic debris and aid in the breakdown of nutrients so that plants may absorb them. Additionally, it strengthens the soil's structure and increases its capacity to retain water, which facilitates the spread of plant roots. We can maintain good soil, which results in stronger plants and more yields, by following these procedures (Carter, 2002).

2.3 Nutrient Management

The goal of managing soil nutrients is to ensure that plants receive the proper quantity of nutrients for growth, without squandering any or endangering the environment. Testing the soil to determine what nutrients it needs and already has is one approach to do this. Farmers can apply the proper kind and quantity of fertilizers based on the results of these tests. This aids in avoiding fertilizer misuse, which can pollute rivers and lakes by washing away fertilizer (Hudson, 1994).

Using cover crops and agricultural rotations is another crucial technique. Diverse plants utilize and incorporate distinct nutrients into the soil. We can maintain the fertility and balance of the soil by rotating the crops planted in a field each year. Grown in between primary crops, cover crops can improve soil fertility and reduce erosion. When combined, these methods save the environment and help preserve healthy soil that may provide good harvests year after year (Ball et al., 2005).

2.4 Water Management

Using water wisely to maintain the health of the soil and plants is known as water management in sustainable soil management. Using effective irrigation techniques, such as drip irrigation, which reduces waste by delivering water straight to the plant roots, is one important strategy. Farmers may use less water while maintaining the proper moisture content in their crops by watering them in this manner (Scavo et al., 2022). Furthermore, methods like mulching—which involves covering the soil with a layer of organic material help hold onto moisture and lessen the need for regular watering. Preventing soil erosion, which may wash away essential nutrients and topsoil, is another crucial component of effective water management. To slow down water flow, farmers can utilize contour ploughing, which follows the natural contours of the soil, and construct terraces on slopes. Maintaining vegetation and planting cover crops alongside streams also aid in holding the soil in place and keep rains from carrying it away. These methods guarantee that water is used prudently and stored for future use, as well as that the soil remains healthy and productive (Qin et al., 2013).

2.5 Erosion Control

In sustainable soil management, erosion control refers to keeping soil from blowing or washing away, and maintaining the productivity and health of the ground. Planting cover crops— plants cultivated expressly to cover the soil is one efficient way. With their roots, these plants help keep the soil in place and shield the top from strong winds and rain. Soil erosion is greatly decreased when the land is covered with plants or mulch. Constructing physical barriers such as windbreaks and vegetation buffers is another method of erosion prevention. Planting rows of trees or bushes along the boundaries of fields creates windbreaks, which help to prevent strong winds from carrying soil away. In order to prevent dirt from washing into streams and rivers, strips of grass or other vegetation are put along waterways called vegetative buffers. By lowering sediment and nutrient discharge, these techniques not only stop soil erosion but also contribute to the preservation of water quality. When combined, these methods guarantee that the soil remains in its proper place, promoting plant development and preserving the ecosystem (McIvor et al., 2014).

2.6 Biodiversity Enhancement

In order to maintain healthy and productive soil, biodiversity augmentation in sustainable soil management refers to expanding the variety of living organisms in and around the soil. One method to do this is to cultivate a variety of crops rather than a single kind. Numerous soil species, including fungi, bacteria, and insects, are supported by this diversity and are crucial to the breakdown of organic matter and the cycling of nutrients. These organisms support plant development by maintaining the soil's fertility and enhancing its structure (Tahat et al., 2020).

Köninger et al. (2021) have reported that using techniques like agroforestry, in which crops are cultivated alongside trees and bushes, is another strategy. Insects and birds that can naturally manage pests find homes in trees. Additionally, they contribute organic matter to the soil through leaf litter and use their roots to stop soil erosion. A diverse range of species that maintain a robust and well-balanced ecosystem are supported when natural habitats surrounding farms, such as wetlands and hedgerows, are preserved. We can cultivate crops in a healthier and more sustainable environment by increasing biodiversity.

2.7 Pollution Prevention

In sustainable soil management, pollution prevention refers to taking action to keep toxic materials out of the soil and adjacent water sources (Dhanalakshmi, 2020). Cutting back on the usage of artificial fertilizers and pesticides is a crucial step in achieving this. Integrated pest management (IPM), on the other hand, employs a variety of techniques to minimize the use of chemicals while controlling pests. For instance, they may grow pest-resistant crops or utilize natural predators to devour the bugs. This contributes to a safer and cleaner environment and soil. Ensuring that hazardous chemicals and industrial trash are not disposed of on farms is another crucial procedure. Frequent soil testing can assist in detecting any pollution in advance, allowing for its correction before it becomes more problematic. We can guarantee pure water, safeguard the health of the soil, and make the environment safer for humans, animals, and plants by preventing pollution (Angon et al., 2023).

2.8 Climate Change Mitigation

Scavo et al. (2022) have reported that reducing greenhouse gas emissions and increasing soil carbon capture are two ways that measures related to climate change mitigation in sustainable soil management assist lessen the effects of climate change. No-till farming, or producing crops without disturbing the soil by ploughing, is one efficient technique. By doing this, more carbon that would otherwise be released into the atmosphere as the greenhouse gas carbon dioxide is retained in the soil. In addition, planting cover crops preserves the soil by absorbing and storing carbon from the atmosphere.

Making the best use of fertilizer is another crucial procedure. A proper fertilizer application both promotes healthy plant growth and inhibits the generation of nitrous oxide, a strong greenhouse gas, from the soil. Agroforestry, or growing trees and bushes next to crops, is another important way to mitigate climate change. Carbon dioxide is taken up by trees from the atmosphere and stored in their roots and wood. Farmers may lessen their total environmental impact and increase the resilience of their farms to climate change by using these techniques (Scavo et al., 2022).

3 Importance of Sustainable Soil Management

We can grow food and protect the environment at the same time when we practise sustainable soil management. Proper care of the soil allows it to remain healthy and support crop growth for an extended period of time. Plants that grow in healthy soil produce better food for humans to consume. Additionally, it aids in environmental protection since soil is home to a variety of microscopic organisms that maintain equilibrium, such as worms and beetles (Pattnaik et al., 2023).

The fact that sustainable soil management aids in mitigating climate change is another important factor. The amount of greenhouse gases in the atmosphere can be decreased by using good soil techniques to store and absorb carbon from the environment. This can help reduce global warming by resulting in cleaner air for us to breathe and less heat being retained in the atmosphere. Therefore, by maintaining healthy soil, we not only assist farmers in producing food, but we also contribute to the planet's long-term health (Pattnaik et al., 2023).

4 Importance of Sustainable Soil Management

4.1 Food Security

Sustainable soil management and food security go hand in hand because healthy soils are necessary to produce enough food to support an expanding population. A consistent supply of food is dependent on the soil being fertile and productive throughout time, which is ensured by sustainable soil management techniques. Farmers may increase soil fertility and lower the likelihood that their crops will fail due to soil degradation by using strategies including crop rotation, cover crops, and the addition of organic matter. Additionally, by strengthening agricultural systems' resistance to environmental shocks like floods, droughts, and extreme weather events—which are occurring more frequently as a result of climate change—sustainable soil management promotes food security. Crops are able to flourish even under unfavorable situations because healthy soils are able to endure these stressors and retain water better. In the end, we can improve food security by putting a priority on sustainable soil management, which will guarantee that soils continue to be productive and able to sustain agriculture for the next generations (Prajapati et al., 2023).

4.2 Environmental Protection

Wolf et al. (2023) studied and reported that healthy soils are necessary for sustaining ecosystems and protecting natural resources, protecting the environment is one of the fundamental objectives of sustainable soil management. We can limit pollution, preserve biodiversity, and lessen soil erosion by putting sustainable methods into practice. By preventing silt and pollutants from washing into rivers and lakes, which can affect aquatic life and ecosystems, erosion prevention contributes to the protection of water quality. Sustainable soil management techniques also support the maintenance of beneficial creatures like fungi, bacteria, and earthworms as well as the preservation of natural environments. These organisms are essential to the cycling of nutrients, the development of soil structure, and plant health. Resilience and general environmental health are enhanced by sustainable soil management, which safeguards these species and their habitats. Ultimately, we can guarantee the long-term viability of agriculture while preserving the health of our planet by giving environmental preservation first priority in soil management.

4.3 Water Quality and Conservation

In order to guarantee that we have clean water to drink and use while also maintaining the health of ecosystems, water quality, and conservation are essential components of sustainable soil management. By lowering the quantity of contaminants that enter bodies of water like rivers, lakes, and streams, sustainable soil management techniques contribute to the improvement of water quality. For instance, we may stop silt and dangerous chemicals from washing into water sources by reducing soil erosion through strategies like growing cover crops and keeping vegetative buffers beside rivers (Ball et al., 2005).

Water conservation is also crucial for managing soil sustainably. Water waste is decreased by using efficient irrigation techniques, such as drip irrigation, which supplies water straight to the plant roots. Mulching is another technique that helps hold onto soil moisture, which lessens the need for regular watering (Ball et al., 2005). We can protect aquatic environments, guarantee that there is enough water for crops to thrive, and save water resources for future generations by practicing responsible water management. In general, water conservation and quality are essential elements of sustainable soil management, supporting agricultural sustainability as well as environmental health.

4.4 Biodiversity Conservation

The goal of biodiversity conservation in sustainable soil management is to preserve the wide variety of organisms that live in the soil. A healthy soil is home to a diverse range of organisms, including bacteria, fungus, worms, and insects, akin to a subterranean metropolis. These microscopic creatures are crucial for recycling nutrients, decomposing dead plant debris, and maintaining the fertility of the soil so that plants can flourish. Farmers may establish habitats that support a wide range of soil life by utilizing sustainable techniques like growing diversified crops and using fewer chemicals (Köninger et al., 2021).

Preserving the diversity of life in the soil benefits not only the insects and worms but also ourselves. Together, these microscopic organisms strengthen the structure of the soil, stop erosion, and even aid in pest management. Growing better crops is only one benefit of maintaining the biodiversity of the soil; another is the preservation of natural ecosystems and the long-term viability of farming. Therefore, maintaining biodiversity is essential to sustainable soil management, which enables us to grow food while maintaining the equilibrium of the soil and the ecosystem.

4.5 Climate Resilience

The term "climate resilience" in sustainable soil management describes a soil's capacity to tolerate and adjust to climatic changes without compromising its production or the health of its ecosystems. In order to protect against extreme weather events like heatwaves, floods, and droughts, healthy soils are essential. They have the capacity to absorb surplus water during periods of high rainfall, minimizing soil erosion and floods, and retain more water during dry periods, lessening the impact of drought on crops. The term "climate resilience" in sustainable soil management describes a soil's capacity to tolerate and adjust to climatic changes without compromising its production or the health of its ecosystems. In order to protect against extreme weather events like heatwaves, floods, and droughts, healthy soils are essential. They have the capacity to absorb surplus water during periods of high rainfall, minimizing soil erosion and floods, and retain more water during dry periods, lessening the impact of drought on crops (Wolf et al., 2023).

Enhancing soil resilience through improved soil structure, increased organic matter content, and biodiversity promotion are some of the practices like conservation tillage, agroforestry, and cover cropping. By improving nutrient cycling, lowering moisture loss, and stabilizing soil temperatures, these techniques assist to increase soil resilience to climatic fluctuations. Farmers may reduce the impact of climate change on agriculture, guarantee food security, and protect the environment for coming generations by implementing sustainable soil management techniques.

4.6 Economic Benefits

Long-term cost savings and increased agricultural output are two ways that sustainable soil management boosts the economy. Healthy soil can result in higher-quality and greater-yield crops, which can boost farmers' income. Farmers may preserve soil fertility without significantly depending on pricey chemical fertilizers and pesticides by employing sustainable techniques like crop rotation and cover crops. Profitability is increased while input expenses are decreased. Additionally, precious farmland is shielded from erosion and deterioration by sustainable soil management, maintaining its long-term worth. Additionally, less water and irrigation are needed on healthy soil, saving farmers money on water bills. Additionally, as customers appreciate food produced responsibly, adopting sustainable techniques may help farmers access new markets and possibilities. All things considered, funding sustainable soil management improves farming enterprises' economic viability while also helping the environment (Prajapati et al., 2023).

5 Strategies for Soil Improvement

Wood and Bowman (2021) reported that healthy soils are essential to sustainable specialty crop production, providing vital nutrients, water, and support for plant growth. Soil-improvement policies aim to increase soil health, fertility, and flexibility to environmental pressures. Cover crops progress soil structure, destroy weeds, and enhance nutrient cycling through biomass combination. Soil fertility, water retention, and biological activity are all improved by raising the amount of organic matter in the soil through agricultural residue management, composting, and green manures (Shinde et al., 2022). Farmers can measure soil quality and make well-informed management decisions by implementing soil health assessment methods and indicators (Guo, 2021). Besides, a very fruitful and promising strategy for improving and remediating soil is the use of nanomaterials (Chen et al., 2022, 2023). In order to protect soil structure and biodiversity, reduced tillage and conservation tillage techniques reduce soil disturbance, erosion, and carbon loss (Kopittke et al., 2019).

There are a lot of obstacles to overcome in order to develop the soil for specialist corps. Precise management and monitoring are necessary to balance nutrient inputs to meet crop demand without negatively affecting the environment. Erosion management methods are necessary because sloped terrain and intense rain events can cause soil erosion. Compaction of soils by heavy machinery and traffic can limit root growth and water infiltration. Growing amounts of nanoplastics may have detrimental effects on soil health and agricultural productivity (Shi et al., 2023). In the interim, creative methods are resolving these issues and developing soil enhancement strategies.

Improved soil fertility, water retention, and nutrient retention are all benefits of using biochar, a stable carbon form made from biomass (Brtnicky et al., 2021). Applying advantageous bacteria progresses soil health overall, disease control, and nutrient availability (Jansson & Hofmockel, 2020, Naz, 2022). Besides, the application of soil amendments such as gypsum, lime, and others rectifies pH imbalances and improves nutrient availability for specialized crops (Bamdad et al., 2022).

The sustainable cultivation of specialty crops is contingent upon effective management of the land and plants. While reducing their detrimental effects on the environment, innovations in precision farming, integrated pest management, and enhancing soil health are increasing productivity. In order to handle problems like pest pressures, labor intensity, and climate change, constant innovation and adaptation are required. By utilizing sustainable practices, biological management methods, and technological advancements, farmers may optimize crop yield and quality while safeguarding the soil's health for upcoming generations. Future opportunities for faster growth in the production of specialized crops will come from ongoing advancements in soil improvement and plant management (Chen & Fang, 2024).

6 Sustainable Development Goals

Based on Bouma and Montanarella (2016) and Keesstra et al. (2016), soil is vital to attaining the United Nations Sustainable Development Goals (SDGs 2), 3, 6, 12, 13, and 15. Crop cultivation is a major source of income for the mainstream individuals living in dearth rural areas. For productivity and income levels in these locations, soil health is a critical element. Soil serves a variety of purposes, including as the foundation for ecosystem services and food production (Bender et al., 2016; Oliver et al., 2019). Furthermore, since lesser crop diseases and pests are associated with soil biodiversity, the ecological services provided by healthy soil systems play a significant role in eradicating hunger and poverty. Regarding the SDGs, soil also influences other significant environmental factors such as water quality and greenhouse gas emissions (Bharati et al., 2002; Franzluebbers, 2005). Figure 3 illustrates an overview of the identified relationships between soil and the relevant SDGs.

It is crucial to provide soil scientific knowledge to practitioners and policymakers who develop and implement SDG programs (see Sect. 3). In order to help produce relevant indicators that are not only backed by science but also helpful to small-scale farmers and other stakeholders, the soil science community must collaborate effectively. A significant focus on accomplishing the Sustainable Development Goals must be placed on scientific research, rather than only on comprehending soil science. The impact of human behavior must be considered in this complex relationship between humans and nature. The impact of socioeconomic activity on soil health must also be considered when conducting sustainability assessments in order to facilitate more informed decision-making (Vogel et al., 2018).

Fig. 3 The significance of soil to the United Nations' Sustainable Development Goals (SDGs)

7 Conclusion

In summary, maintaining the productivity of our farms, the health of the environment, and the welfare of future generations all depend on sustainable soil management. It is possible to maintain healthy soils that sustain thriving ecosystems and sustainable food production by implementing measures that improve nutrient management, protect soil structure, reduce erosion, and sustain biodiversity. In addition to helping farmers by increasing agricultural yields and cutting expenses, these initiatives also support more general environmental objectives including preventing climate change and preserving water quality. Adopting sustainable soil management techniques is essential to creating a resilient and sustainable future for people and the environment.

References

Angon, P. B., Mondal, S., Jahan, I., Datto, M., Antu, U. B., Ayshi, F. J., & Islam, M. S. (2023). Integrated pest management (IPM) in agriculture and its role in maintaining ecological balance and biodiversity. *Advances in Agriculture, 2023*(1), 5546373. https://doi.org/10.1155/2023/5546373

Ball, B. C., Bingham, I., Rees, R. M., Watson, C. A., & Litterick, A. (2005). The role of crop rotation in determining soil structure and crop growth conditions. *Canadian Journal of Soil Science, 85*, 557–577. https://doi.org/10.4141/S04-078

Bamdad, H., Papari, S., Lazarovits, G., & Berruti, F. (2022). Soil amendments for sustainable agriculture: Microbial organic fertilizers. *Soil Use and Management, 38*, 94–120.

Bender, S. F., Wagg, C., & van der Heijden, M. G. (2016). An underground revolution: Biodiversity and soil ecological engineering for agricultural sustainability. *Trends in Ecology & Evolution, 31*, 440–452.

Bharati, L., Lee, T. H., Isenhart, T., & Schultz, R. (2002). Soil-water infiltration under crops, pasture, and established riparian buffer in Midwestern USA. *Agroforestry Systems, 56*, 249–257.

Bouma and Montanarella. (2016). Facing policy challenges with inter- and transdisciplinary soil research focused on the UN Sustainable Development Goals. *The Soil, 2*, 135.

Brtnicky, M., Datta, R., Holatko, J., Bielska, L., Gusiatin, Z. M., Kucerik, J., Hammerschmiedt, T., Danish, S., Radziemska, M., Mravcova, L., Fahad, S., Kintl, A., Sudoma, M., Ahmed, N., & Pecina, V. (2021). A critical review of the possible adverse effects of biochar in the soil environment. *Science of the Total Environment, 796*, 148756, ISSN 0048-9697. https://doi.org/10.1016/j.scitotenv.2021.148756

Carter, M. R. (2002). Soil quality for sustainable land management. Organic Matter and aggregation interactions that maintain soil function. *Agronomy Journal, 94*, 38–47. https://doi.org/10.2134/agronj2002.3800

Chen, X., Chu, S., Chi, Y., Wang, J., Wang, R., You, Y., Hayat, K., Khalid, M., Zhang, D., & Zhou, P. (2023). Unraveling the role of multi-walled carbon nanotubes in a corn-soil system: Plant growth, oxidative stress and heavy metal (loid) s behavior. *Plant Physiology and Biochemistry, 200*, Article 107802.

Chen, X., & Fang, L. (2024). Plant management and soil improvement in specialty crop production. *Applied Sciences, 14*(13), 5915. https://doi.org/10.3390/app14135915

Chen, X., Wang, J., Wang, R., Zhang, D., Chu, S., Yang, X., Hayat, K., Fan, Z., Cao, X., & Ok, Y. S. (2022). Insights into growth-promoting effect of nanomaterials: Using transcriptomics and metabolomics to reveal the molecular mechanisms of MWCNTs in enhancing hyperaccumulator under heavy metal (loid) s stress. *Journal of Hazardous Materials, 439*, 129640.

Dhanalakshmi, T. (2020). Do we live for microorganisms or do they live for us? In *Microbial biodiversity* (p. 14). Cambridge Scholars Publication, PISBN (13): 978-1-5275-4818-3.

El-Beltagi, H. S., Basit, A., Mohamed, H. I., Ali, I., Ullah, S., Kamel, E. A. R., Shalaby, T. A., Ramadan, K. M. A., Alkhateeb, A. A., Ghazzawy, H. S. (2022). Mulching as a sustainable water and soil saving practice in agriculture: A review. *Agronomy, 12*, 1881. https://doi.org/10.3390/agronomy12081881

Franzluebbers, A. J. (2005). Soil organic carbon sequestration and agricultural greenhouse gas emissions in the Southeastern USA. *Soil and Tillage Research, 83*, 120–147.

Guo, M. (2021). Soil health assessment and management: Recent development in science and practices. *Soil Systems, 5*, 61.

Hudson, B. D. (1994). Soil organic matter and available water capacity. *Journal of Soil and Water Conservation, 49*, 189–194. https://doi.org/10.1201/9780429445552-36

Idowu, J., Ghimire, R., Flynn, R., & Ganguli, A. (2020). *Soil health-importance, assessment, and management*. Cooperative Extension Service, Circular 694B

Jansson, J. K., & Hofmockel, K. S. (2020). Soil microbiomes and climate change. *Nature Reviews Microbiology, 18*, 35–46.

Keesstra, S. D., Bouma, J., Wallinga, J., Tittonell, P., Smith, P., & Cerdà, A. (2016). The significance of soils and soil science towards realization of the United Nations Sustainable Development Goals. *The Soil, 2*, 111–128. https://doi.org/10.5194/soil-2-111-2016

Köninger, J., Lugato, E., Panagos, P., Kochupillai, M., Orgiazzi, A., & Briones, M. J. (2021). Manure management and soil biodiversity: Towards more sustainable food systems in the EU. *Agricultural Systems, 194*, 103251. ISSN 0308-21X, https://doi.org/10.1016/j.agsy.2021.103251

Kopittke, P. M., Menzies, N. W., Wang, P., McKenna, B. A., & Lombi, E. (2019). Soil and the intensification of agriculture for global food security. *Environment International, 132*, Article 105078.

McIvor, I., Youjun H., Daoping, L., Eyles, G., & Pu, Z. (2014). Agroforestry: Conservation trees and erosion prevention. In N. K. Van Alfen (Ed.), *Encyclopedia of agriculture and food systems* (pp. 208–221). Academic Press. ISBN 9780080931395, https://doi.org/10.1016/B978-0-444-52512-3.00247-3

Naz, M. (2022). The soil pH and heavy metals revealed their impact on soil microbial community. *Journal of Environmental Management, 321*, 115770.

Oliver, D. P., Li, Y., Orr, R., Nelson, P., Barnes, M., & McLaughlin, M. J. (2019). The role of surface charge and pH changes in tropical soils on sorption behaviour of per- and polyfluoroalkyl substances (PFASs). *Science of the Total Environment, 673*, 197–206.

Pattnaik, B. K., Sahu, C., Choudhury, S., Santra, S. C., & Moulick, D. (2023). Importance of soil management in sustainable agriculture. In M. Hasanuzzaman (Eds.), *Climate-resilient agriculture* (vol 1). Springer. https://doi.org/10.1007/978-3-031-37424-1_22

Prajapati, S. K., Verma, S., & Yadav, P. K. (2023). Agricultural transformation through precision farming: Using agrochemicals efficiently to enhance productivity. *Food and Scientific Reports, 4*(5), 36–50.

Pretty, J. (2008). Agricultural sustainability: Concepts, principles and evidence. *Philosophical Transactions or the Royal Society b., 363*, 447–465. https://doi.org/10.1098/rstb.2007.2163

Qin, W., Chi, B., & Oenema, O. (2013). Long-term monitoring of rainfed wheat yield and soil water at the loess plateau reveals low water use efficiency. *PLoS ONE, 8*, Article e78828. https://doi.org/10.1371/journal.pone.0078828

Scavo, A., Fontanazza, S., & Restuccia, A. (2022). The role of cover crops in improving soil fertility and plant nutritional status in temperate climates. A review. *Agronomy for Sustainable Development, 42*, 93. https://doi.org/10.1007/s13593-022-00825-0

Sharma, P., Singh, A., Kahlon, C., Brar, A., Grover, K., Dia, M., & Steiner, R. (2018). The role of cover crops towards sustainable soil health and agriculture—A review paper. *American Journal of Plant Sciences, 9*, 1935–1951. https://doi.org/10.4236/ajps.2018.99140

Shi, Q., Li, Z., Xu, Y., Yan, T., & Chen, M. (2023). Correction: Shi et al. Dynamic Scenario Simulations of Sustainable Rural and Towns Development in China: The Case of Wujiang District. *Sustainability* 2023, *15*, 8200. *Sustainability, 15*, 14254. https://doi.org/10.3390/su151914254

Shinde, R., Shahi, D. K., Mahapatra, P., Singh, C. S., Naik, S. K., Thombare, N., & Singh, A. K. (2022). Management of crop residues with special reference to the on-farm utilization methods: A review. *Industrial Crops and Products, 181*, Article 114772.

Tahat, M. M., Alananbeh, K. M., Othman, Y. A., & Leskovar, D. I. (2020). Soil health and sustainable agriculture. *Sustainability, 12*(12), 4859. https://doi.org/10.3390/su12124859

Tracy, S. R., Black, C. R., Roberts, J. A., & Mooney, S. J. (2011). Soil compaction: A review of past and present techniques for investigating effects on root growth. *Journal of the Science of Food and Agriculture, 91*, 1528–1537.

USDA. Natural resources conservation service soil health portal. https://www.nrcs.usda.gov/wps/portal/nrcs/main/soils/health/

Vogel, H. J., Bartke, S., Daedlow, K., Helming, K., Kögel-Knabner, I., & Lang, B. (2018). A systemic approach for modeling soil functions. *The Soil, 4*, 83.

Wolf, M. K., Wiesmeier, M., Macholdt, J. (2023). Importance of soil fertility for climate-resilient cropping systems: The farmer's perspective. *Soil Security, 13*, 100119. https://doi.org/10.1016/j.soisec.2023.100119. ISSN 2667-0062.

Wood, S. A., & Bowman, M. (2021). Large-scale farmer-led experiment demonstrates positive impact of cover crops on multiple soil health indicators. *Nature Food, 2*, 97–103.

Soil Restoration: Methods and Challenges

Ajay Bhalerao and Ashish Jain

Abstract

Soil restoration brings degraded soils back to their original state and has an influence on crop productivity through biological management. Soil restoration plays a key role in managing land, but most limits have room to get better. We need to use long-lasting ways to take care of the land. This helps stop soil from washing away, keeps farms productive, and protects plants and animals over time. The big issues that come from worn-out soil should push us to study and make better ways to fix damaged ground. This chapter looks at things that might hold back soil recovery. Agroforestry, agricultural conservation programs, and soil research are reviewed as ways to enhance soil health. The study continues to insist on the holistic approach to soil biology, chemistry, and physics as the solution to challenges such as the requirement to understand sensitivities concerning soil health; shortage in funding; and resistance from creditors on project interventions. This implies the role of financial inputs and adaptive management in keeping the long-term viability of soil restoration projects. This literature underlines the consideration for unique local factors and contextual factors while designing effective strategies aimed at restoring soil health- context-specific approaches are to be adopted.

Keywords

Soil restoration · Soil health · Sustainable land management · Soil fertility

1 Introduction

The earth is a wonderful natural system that provides humans with essential blessings and supports a sweet and dignified life. The land is especially important for agriculture, feeding a growing population, protecting ecosystems, providing water, vitamins, and other essential nutrients; cycling; managing carbon emissions; and climate change control. It is seen as an economically beneficial resource for human betterment and has social, bio-cultural, and non-secular importance (Abhilash, 2021). One of the most emerging issues of our day is soil degradation. A large part of the regions of cultivable soil has been degraded due to land grabbing, erosion, and urbanization. More than a third of the ground soil is ideal to be degraded these days; this is a territory wherein biodiversity has decreased, soil fertility has been weakened, and the soil is a way extra liable to erode or become desertified. Thus, the hunt is on for practical methods to refill part of our degraded land with nutrient-wealthy topsoil (Abhilash, 2021).

Research on soil remediation techniques and their challenges has extensively explored various strategies, including biotech and traditional farming methods. Its consistent intention is to find solutions that aid progressed soil fitness, increase ecosystem resilience, and preserve land management practices. Several soil healing strategies stress the advantages of soil amendments (along with recovery-minded organic count number additions) and the function that conservation agriculture and agroforestry play in enhancing soil fertility and physical fine (soil shape) in many cropping structures. The research studies of those evaluations spotlight the importance

A. Bhalerao
Microbiology and Biotechnology Department, BNN College, Dhamankar Naka, Bhiwandi, Maharashtra, India

A. Jain (✉)
Department of Microbiology, Smt. C.H.M College, Ulhasnagar, India
e-mail: ajcdri@gmail.com

University of Mumbai, Maharashtra, India

J. A. Parray (ed.), *Soil and Land Use Change*, Sustainable Landscape Planning and Natural Resources Management,
https://doi.org/10.1007/978-3-031-93075-1_5

of controlling soil and water loss through agronomic practices, diversifying vegetation (planting a much wider range of plants), and managing natural counts to repair the energy (productivity) and resilience (capability to resist and recover from pressure or damage) of soil assets. For instance, research in Missouri, USA, demonstrates that these structures enhance soil fertility and increase water retention. Similarly, there has been more research that explores the effectiveness of soil bioengineering measures in preserving slope formalism and in landslide prevention in mountainous regions (Udawatta et al., 2021, Lu et al., 2019).

Even with continuing research, many more obstacles remain on the path to widespread and general success for soil reclamation. It is crucial to take into account the particular issues and experiences of such land reclamation projects because similar procedures may frequently be modified to fit different local conditions. For example, a case study in Missouri, USA, shows that agroforestry systems increase the fertility of soil and water retention (Udawatta et al., 2021). After recently researching the feasibility of using soil bioengineering techniques to stabilize slopes and prevent landslides in mountainous areas is investigated. But even though there is increasing research on this topic, major obstacles to the practical application and future development of landfilling still exist (Lu et al., 2019).

2 Techniques for Soil Restoration

2.1 Conservation Agriculture

The utilization of the soil in a sustainable fashion through agriculture is a meaningful solution to soil preservation. The essence is the employment of approaches that not only improve the yield of production but also save and regenerate the soil and add value to livestock. One of the main techniques involves "minimal tillage" which refers to preserving the structure of the soil with minimum disturbance. This involves approaches such as reduced or zero tillage when preparing land for planting crops. The organic matter content in the soil, soil structure, and microbial population could be conserved by these methods (Li et al., 2021, Bekele, 2020). Among other advantages over the traditional methods of tilling, this strategy has been proven to be efficient in improving the soil systems, reducing the amount of erosion, conserving water, and releasing less carbon dioxide into the environment. Reduced tillage improves soil organic carbon content, water retention, and soil aggregation (Li et al., 2021). Bekele (2020) found that limiting tillage to reasonable boundaries is necessary to limit soil erosion, reduce moisture loss, and boost the soil's biodiversity. The study further demonstrated the benefits of low-tillage farming through experiments conducted in various locations after conversion to farmlands. Adam and Abdulai (2023) discovered that minimum tillage across the northern Savannah zone in Ghana elevated crop yield, soil structure, and water penetration.

Crop rotation is one of the most suitable methods of farming. Crop rotation involves farmers planting incorporated crops systematically in predetermined pieces of farmland for an agreed duration. This system provides farmers with numerous benefits, including the disruption of pest and disease cycles. Additionally, this system aids in reducing the use of chemical pesticides, which in turn reduces the number of pests, as different crops tend to attract different types. Therefore, no single pest can be present in plants growing on the rotating fields. This eliminates the need for chemical pesticides. This will assist the farmer in avoiding the expenses associated with purchasing and labor-intensive chemical applications. Since the nutrient requirements of different crops vary, this approach also helps to prevent the depletion of specific nutrients in the soil. Similarly, some crops can fix nitrogen from the atmosphere and support soil health. Sustained cropping makes the pests, diseases, and weeds specific to certain crops (Nadi, 2023; Zhao et al., 2020; Shah et al., 2021).

Cropping has been proven to decrease the amount of various nutrients required for crop growth and the degradation of organic matter in the soil by microbiological and other beneficial soil organisms after being cultivated many times. However, by changing the plant variety and planting order, crop rotation contributes to the long-term sustainability of agriculture (Nadi, 2023; Zhao et al., 2020; Shah et al., 2021).

One study highlights the positive impacts that crop rotation generates in improving the nutrient recycling process, weed management, as well as reduction in soil pathogens (Zhao et al., 2020). Research provides us with the details we need about the agronomic benefits of different crop rotation systems like how they help to increase resistance against climatic changes, hence reducing susceptibility to diseases from insects thereby ensuring there is less pest population as well as quality agriculture soils for sustainability purposes (Shah et al., 2021). Examples from various regions show that crop rotation can enhance productivity, improve soil fertility as well as decrease dependence on external inputs. The three specific crop rotation schemes include barley-mungbean (BM), wheat-mungbean (WM), and canola-mungbean (CM). Material inputs of gasoline and chemical fertilizers were the most important factors affecting energy use and greenhouse gas emissions in all three agroecosystems (Nadi, 2023).

When talking about conservation agriculture basic cover crops are essentially meant for soil preservation, reducing erosion of the landscape, and increasing soil organic matter. They are non-profit making types, best utilized in minimizing soil losses, especially through run-off, building up the fertility of the field, and inhibiting weeds, whenever main economic crops are not in season. Besides promoting fertility in the soil, the crops promote healthy agricultural ground by adding organic matter and preventing nutrient losses through leaching or run-off; thus leading to sustainable farming practices. This proves the importance of cover crops to improve the strength of resilience of an agroecosystem, raise the level of biodiversity, and minimize the ecological impact of conventional agriculture. Soil fertility, water retention, and nutrient recycling are some of the reasons why cover crop usage is highly recommended (Wood & Bowman, 2021).

In summary, conservation agriculture employs several techniques to boost soil fertility and enhance food productivity without damaging the land. These techniques include reduced tillage, crop rotation, and cover cropping, which aim to improve soil quality, sustain resources, and strengthen agriculture. However, for conservation agriculture to be effective, it must be specific to a particular region, align with farmers' preferences, and align with their economic circumstances.

2.2 Soil Amendments

Soil amendments improve fertility, management, and availability of nutrients for soil restoration and as such they are vital for restoring soil productivity. The impact of various soil amendments as demonstrated by studies and scientific works was analyzed.

(1) Organic Matter Addition

The addition of organic substances enhances the nutrient content and physical qualities of the ground by mixing it with plant residues and other organic matter, thereby improving its fertility and texture. Various research studies have testified to the positive impacts of incorporating organic materials with soil as far as nutrient availability, water retention, and microbial activities that enhance its quality are concerned (Gunina & Kuzyakov, 2022). This technology not only enhances crop yields; it also promotes environmental practices (which lessen the use of chemical fertilizers) and reduces greenhouse gas emissions in agriculture. By incorporating natural cloth with manure or wood ash into the soil, several upgrades (closer to greater soil carbon storage) can be made, because accelerated microbial activity levels have been validated using examples of compost and biochar. However, this approach requires careful management, as the balance between organic inputs and soil health is crucial. Although the benefits are significant, it is important to consider the local context when implementing these practices.

The investigations emphasize the significance of the organic matter in the process to improve soil consistency, conserve organic moisture, and provide nutrients, while also advising the usage of environmentally friendly organic inputs to avoid environmental degradation. There are numerous motives why the inclusion of natural fabric is useful. For instance, making use of manure increases fertilizer efficiency, reduces nutrient run-off losses, and complements soil's natural carbon levels (Gross & Glaser, 2021).

(2) Nutrient Balancing

Proper nutrient balance is necessary to maximize plant growth and productivity while minimizing harm to the environment. Application of fertilizers for nutrient additions should be done progressively to avoid excesses that maintain soil fertility within appropriate thresholds supporting crops against diseases caused by insects or plants themselves (vines). Optimum levels of soil nutrients should always be maintained by being careful when applying fertilizers because this would ensure healthy plant growth with high crop yields and fewer negative effects on the environment. Research suggests that farmers can optimize fertilizer use, reduce nutrient run-off, and improve crop nutrient uptake performance by means of exactly controlling nutrients based on soil evaluation and crop requirements. By carefully distributing nutrients to meet plant requirements and minimizing nutrient imbalances and their outcomes on the environment, this approach helps sustainable farming methods. The investigation explores the role that nutrient control performs in improving crop output, soil fertility, and nutrient utilization efficiency (Lu et al., 2019).

Through the technology, farmers are now able to customize nutrient dolls based on soil examination results and crops' needs, hence growth and fertilizer utility is enhanced. Besides that, to the health companies, proving to be effective the interventions related to lower levels of sugar, salt, and saturated fats and higher levels of vegetables, fruits, and whole grains can be done. This ensures the embracement of ecological farming through the feeding of the plants with nutrients which calls for the coupling of wishes, rustification of imbalances, and neutralization of environmental effects.

2.3 Agroforestry

The systems empower the stability of ecosystems, enrich soil quality, and increase biota by combining plant, crop, or animal growth on the same land.

(1) Tree integration: This process known as tree integration designates the combination of trees with the urban and rural environments enhances sustainability, environmental vitality, and more. As a result, productivity is extended and resistance to environmental troubles like erosion and drought is elevated. Agroforestry and sustainable land control principles are accompanied by using tree

Fig. 1 A linked concept between agroforestry and soil quality, health, and fertility

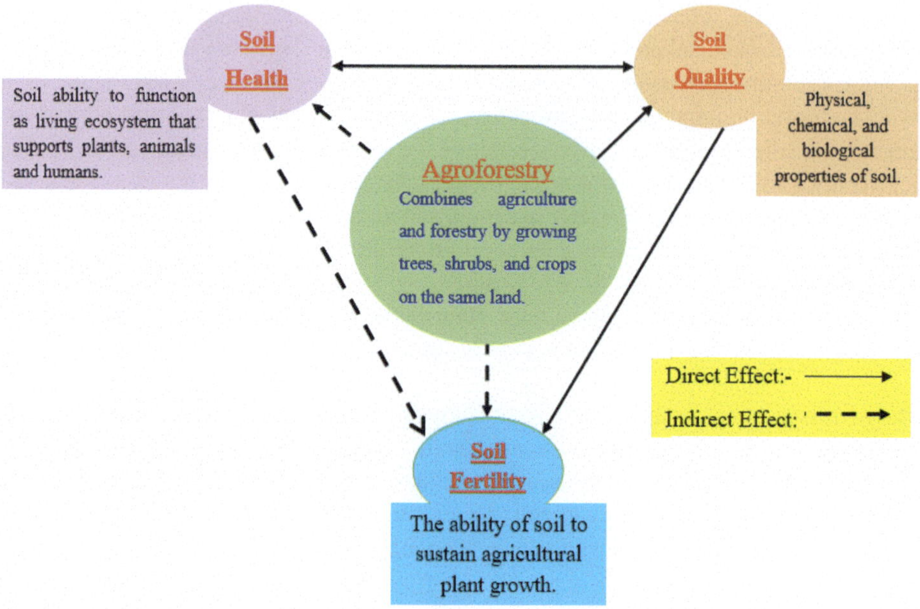

integration, which has numerous blessings for both atmosphere health and human well-being (Brancalion & Holl, 2020) (Fig. 1).

Trees help improve soil fertility through nitrogen fixation, cycling of nutrients, and accumulation of organic matter. Tree integration is very important in agroforestry systems (Lelamo, 2021). There are numerous ecological roles that timber plays, which include controlling erosion, regulating water, and imparting habitat for organisms that can be treasured by the environment. Numerous agroecological zones show how integrating trees improves soil health and yields in agriculture. For instance, tests from suggested improvements in agricultural yields, water infiltration, and soil fertility in agroforestry structures set up in Gowa, Indonesia (Graham et al., 2022).

(2) Increasing Biodiversity: Efforts should be designed in a strategic way to increase the number and variety that exist within any given area. Among other strategies, including the planting of shrubs, formation of herbaceous corridors, and adoption of agroforestry can efficiently improve biodiversity through the provision of some habitats as well as unique species. Research performed in plenty of ecosystems the world over has shown that those strategies increase atmosphere services, improve ecosystem resilience, and help long-term ecological balance. Agroforestry, especially concerning family gardens–plays in preserving biodiversity and enhancing environmental resilience (Yashmita-Ulman et al., 2021).

Structural complexity, choice of tree species, and connections within the landscape have been pointed out as important factors that promote biodiversity in agricultural systems. There is ample evidence from various parts of the world including developed and developing countries showing the positive effects of biodiversity through the establishment of agroforestry systems (Udawatta et al., 2021). Some examples of these are; there was a research study into tree planting techniques utilized by the small-scale farmers in Mt. Elgon region within Uganda. The primary objective of this research was to get an understanding of why farmers prefer planting exotics over indigenous tree species despite some factors that dictate this choice such as price differences among trees on sale, while also looking at reasons them keeping indigenous trees on their land. It determined their ambitions about increasing their tree stockpile with Indigenous ones and the hindrances they met along this path (Graham et al., 2022).

2.4 Soil Bioengineering

The innovative method of using plants and natural materials to stabilize slopes and fight against erosion is employed here. It is the blend of biology with engineering. It offers inexpensive environmentally attractive solutions to locations that are prone to erosions by utilizing the unique abilities of plants in terms of soil interaction.

(1) Vegetation Stabilization: Vegetation stabilization entails the use of flowers mainly long-root species to stabilize soil and save erosion. Research carried out on the Chinese Loess Plateau demonstrates how plants make use of approaches that include root reinforcement for strengthening soil shape, reducing slope instabilities, and enhancing slope stability (Yue et al., 2024).

Among many fields of science, the place to do research is chosen and some studies focus on species with requirements such as surface coverage, rate of root development, and adaptation to conditions of the site for use of vegetative technique in stabilization. They emphasized the significance choosing of certain species (based on their adaptation, growth patterns, and root system) by choosing good plant species for a particular place (Archibong et al., 2020). Various research from extraordinary areas entails how powerful plant stabilization methods are for soil bioengineering projects. This would be, for example, an examination of the distinctive plant regrowth techniques and their influence on Yanchi County's ability to curtail wind erosion. The consequences inculcate a disinterested usage of ecological records of the operations restoration and the rise of ecological conservation policies, helping the authorities with the policies selection using empirical records basis (Luo et al., 2020).

(2) Soil erosion management: Soil bioengineering methods for soil erosion control additionally make use of natural manipulation systems together with promoting plant life to prevent erosion. It has been mentioned that techniques which include brush layering and live fascine creation can save soil erosion and loss on the terrain. Various investigations reveal to prevent soil erosion and additionally moderate the ecosystem (Gonzalez-Ollauri et al., 2022). Soil bioengineering practices have enabled successful erosion control projects in coastal zones, hillslopes, and riverbanks. For instance, research indicates how bioengineering answers, inclusive of stay crib walls and brush beds, can also stabilize riverbanks and lower sediment transport fees.

2.5 Soil Bioremediation

By the usage of the innate capacities of microbes, plants, and their enzymes to break down or detoxify pollution, soil bioremediation offers a sustainable and environmentally beneficial approach to soil recovery. This element will test how powerful soil bioremediation strategies.

(1) Microbial Degradation: This is a system of breaking down pollution via microbes into less dangerous or negative forms. It is found that the various metabolic capacities of soil microorganisms which are supplemented with soil infected with PAHs in breaking down a huge type of natural contaminants, at the side of pesticides and hydrocarbons (Lu et al., 2019).

There are several microbial degradation mechanisms and their applications in soil remediation. Understanding the dynamics of networks and microbial ecology is emphasized as essential to enhance degradation rates and optimize bioremediation strategies (Bhagat & Jain, 2024). Examples from actual sites show how decomposition may effectively reduce soil contamination. For example, studies display that extended publicity to aromatic chemical compounds promotes the usage of polycyclic aromatic hydrocarbons (PAHs) as a supply of carbon and electricity by using bacterial strains. The consequences of the research display that the overall performance of the hydrocarbon biodegradation method is no longer extra fantastic with the useful resource of using surfactants (Smułek et al., 2020).

(2) Plant-Based Remediation: Plants, via techniques such as extraction, rhizodegradation, and phytostabilization, can effectively decontaminate soil (Bhagat et al., 2023, 2024). Many such techniques are included in the umbrella term phytoremediation, which refers to the use of natural vegetation for remediating soils contaminated by complex organic or inorganic pollutants. A recent study in 2022 found that certain plants near Iran's Khatunabad Copper Smelter and Refinery can effectively remove metals from soils, offering a cost-effective remediation solution (Siyar et al., 2022). The comparative study will explore mechanisms and practical applications of phytoremediation technologies for the treatment of soil pollution (Fig. 2). It is found that the selection of various plant species and growth conditions optimization enhances the overall efficiency of remediation programs. The efficacy of plant-based total remediation in several soil infections is carried out globally by phytoremediation programs. For example, researchers demonstrate the powerful utilization of the Eucalyptus globulus plant to phytoextract heavy metals and natural contaminants from polluted Moroccan iron mine soil (Fig. 3) (El Rasafi et al., 2021).

2.6 Water Management

By stabilizing moisture levels, reducing erosion, and boosting agricultural output, powerful water manipulation is critical to regaining the health of the soil. This phase examines the techniques and practices of water management based totally on arguments, instances, and investigations.

(1) Watering Methods: To keep a good enough stage of soil moisture and save needless wastage, quality watering strategies must be applied. To maximize water transport, reduce run-off, and shop water resources, they strain the need for suitable timing, irrigation scheduling, and irrigation system layout (Luo et al., 2020). The benefits of the usage of powerful irrigation strategies are illustrated by numerous agroecological zones. For example, in the Campania Felix location of southern Italy, the study

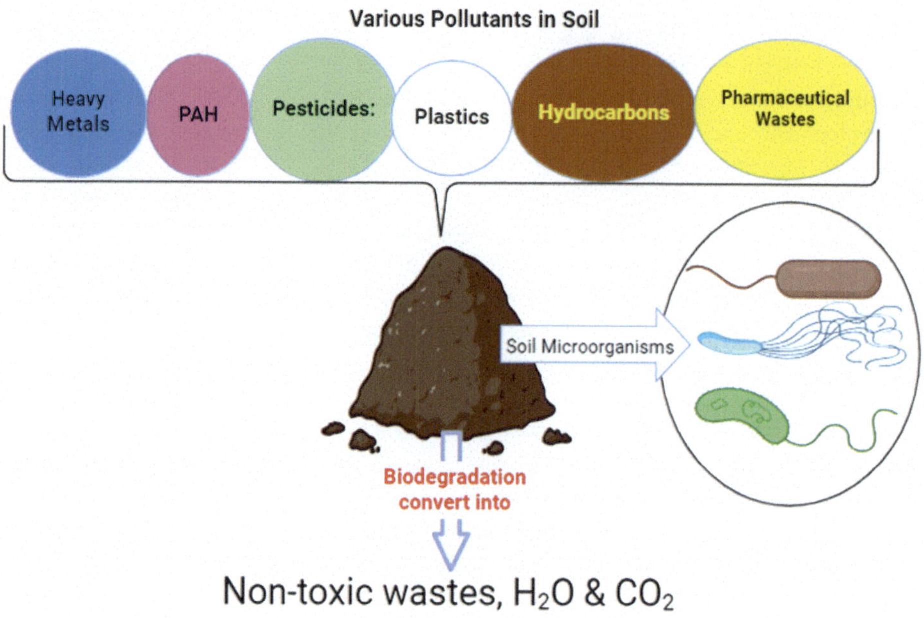

Fig. 2 Bioremediation by figure soil microbes

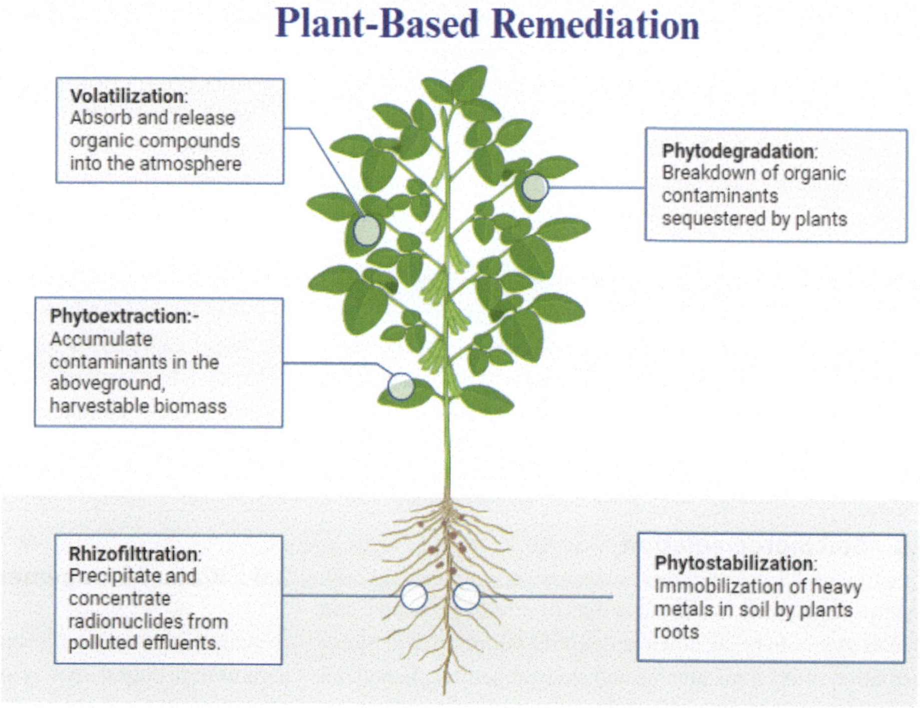

Fig. 3 Schematic representation of different phytoremediation approaches by plants under study

suggests the useful influences of irrigation help gadgets, i.e. LCIS DSS, on soil moisture retention, crop improvement, maize output, and water conservation (Bonfante et al., 2019).

(2) Rainwater Harvesting: Where water delivery is unreliable or scarce, numerous methods of rainwater harvesting which include check dams, contour bunds, rooftop rainwater harvesting structures, etc. are used to deliver it sustainable agriculture has been completed, soil moisture has been accelerated and rainwater harvested for agricultural purposes. To optimize water seizure and decrease soil erosion, they stress the need for web page-particular planning, suitable design, and protection (Zhao et al., 2020).

The usefulness of water-accumulating strategies in raising soil moisture ranges and boosting agricultural output is established by research from water-pressured areas. For instance, studies show that Nature-Based Solutions (NBS) offer advantages over traditional procedures. These advantages consist of more water great, improved biodiversity, social co-blessings, and decreased strength utilization due to progress in indoor climate, and advanced urban microclimate. Based on these consequences, the studies recommend giving systemic change towards NBS precedence over traditional water infrastructure (Oral et al., 2020).

3 Challenges in Soil Restoration

Difficulties in restoring soil achieving prolonged-time period sustainability in soil restoration obligations requires overcoming numerous demanding situations related to protection and ensuring ongoing success. Through an evaluation of soil recovery tasks, this segment will investigate the problems.

A. Resistance to Change: One of the most important boundaries to soil recovery is stakeholder resistance to trade, especially among farmers and landowners. Farmers are regularly reluctant to implement new soil conservation techniques due to deeply ingrained cultural norms, a loss of records, or a notion of economic danger (Chinseu et al., 2019). This difficulty comes from exclusive agricultural places, which spotlight the want to address sociocultural variables and present incentives to promote the adoption of sustainable soil control techniques. Initiatives such as information-sharing and farmer engagement, for example, were effective in overcoming competition to change (Willson et al., 2021). Adopting sustainable procedures for soil control is hampered by stakeholders' resistance to change, together with farmers and landowners. The necessity for custom-designed techniques that deal with sociocultural components and provide incentives to promote behavioral shifts in the direction of extra-sustainable land management practices (Adam & Abdulai, 2023; Willson et al., 2021).
B. Financial Restraints: Specifically, financial constraints and insufficient capital are the primary obstacles that prevent grassroots farmers and small depots from implementing soil recovery strategies. Economic hindrances exacerbate the task of soil restoration, particularly for small-scale farmers and underprivileged communities. We need to make financial assistance available to support these individuals and ensure that their past accomplishments in this area are not lost. We need sufficient funding to direct investments in soil conservation strategies, such as selecting systems, implementing innovative technologies, or gaining access to organic amendments.
C. Policy and Governance: Soil degradation is crucial. Historically, a multitude of obstacles have impeded the development of effective frameworks and corresponding policies, which are essential for successful soil restoration. Institutional frameworks such as agricultural standards and advisory directories on soil maintenance must be set up as well as land utilization initiatives (Löbmann et al., 2022). The effort to minimize soil degradation will demand an inclusive regulation that combines various environmental regulations, land tenure, and farming subsidies.
D. Awareness and Knowledge: Farmers are unable to extensively employ soil restoration techniques due to a lack of awareness and insufficient knowledge of the technologies involved in restoring soil health. Such technologies should be promoted very much through training and extension programmes to educate farmers, decision-makers, and also the public on how one can restore the soil and fertility of their land in an environmentally friendly manner. Concerning research in 2021, within Turkey, it is necessary to strengthen the attitude towards the broadcasting of information relating to the restoration of soils as well as the sustainable ways of farming. In addition, research shows how essential network-based projects and educational activities have been in selling grassroots involvement and cognizance. Soil recovery efforts that endure forever depend on a continuous commitment to preservation, adaptable leadership, and institutional support (Shah et al., 2021). Various examples from community-based total absolute responsibilities are used to demonstrate the value of training and extension programs in promoting soil conservation measures. For example, responsibilities that provide training, demonstration plots, and extension services were successful in building community capacity and promoting network participation in soil recovery projects (Johnson et al., 2019).
E. Sustaining and Extended Sustainability: To keep this long-term contribution by soil maintenance work, maintenance-related issues need to be dealt with and continued "fitness for use" needs to be assured. Soil fitness as well as production can be maintained by "maintenance practices." This includes "maintaining cover crops, preventing erosion, and adding more biotic elements." Furthermore, adaptive control methods that can adapt to shifting environmental situations and transferring socio-economic settings are necessary to guarantee the long-term viability of soil recovery initiatives (Moyo et al., 2021). Maintaining soil recuperation efforts and making sure of their sustainability through the years also requires securing sufficient institutional and monetary guides (Abhilash, 2021).

4 Conclusion

A variety of challenges, improvements, and possibilities for destiny are shining within the soil of healing communication. In particular, smallholder farms and economically disadvantaged communities require tailored approaches to soil restoration that take sociocultural aspects into account. Effective implementation of such projects critically depends on stakeholder engagement, which creates mutual understanding and support among all parties concerned. The long-term success of soil restoration programmes depends on the supply of financial support, such as advancing agroforestry, soil supplementation, and conservation agriculture due to constraints towards improving soil resilience and fertility. Soil restoration initiatives, being adaptive management techniques, will work out for the long-term results because they will result in continuous learning and adaptation to the changing socio-economic and environmental conditions. If soil restoration efforts are to be successful, then certainly, inadequate policy support and governance structures will have to be addressed. Sustainable soil restoration calls for policy measures that can integrate land tenure, agricultural subsidies, and environmental legislation. This paper, therefore, underscores the need for teamwork, creativity, and sustainability in stopping soil degradation and keeping the soil productive for the coming generations. As the world grapples with problems of economic inequality and environmental degradation, it is clear that comprehensive and context-specific solutions are needed for sustainable development.

References

Abhilash, P. C. (2021). Restoring the unrestored: Strategies for restoring global land during the UN decade on ecosystem restoration (UN-DER). *Land, 10*(2), 201. https://doi.org/10.3390/land10020201

Adam, B., & Abdulai, A. (2023). Minimum tillage as climate-smart agriculture practice and its impact on food and nutrition security. *PLoS ONE, 18*(12), Article e0287441. https://doi.org/10.1371/journal.pone.0287441

Archibong, G. A., Sunday, E. U., Akudike, J., & Okeke, O. (2020). A review of the principles and methods of soil stabilization. *International Journal of Advanced Academic Research, 6*(3), 89–115. https://www.researchgate.net/publication/342448451

Bekele, D. (2020). The effect of tillage on soil moisture conservation: A review. *International Journal of Research Studies in Agricultural Sciences, 6*(10). https://doi.org/10.20431/2454-6224.0610004

Bhagat, S., & Jain, A. (2024). Phosphate-solubilizing bacteria-assisted phytoremediation of metalliferous soils. In *Microbiome-assisted bioremediation* (pp. 327–348). Academic Press. https://doi.org/10.1016/B978-0-443-21911-5.00008-8

Bhagat, S., Shete, P., & Jain, A. (2023). Plant health: feedback effect of root exudates and rhizobiome interactions. In *Rhizobiome* (pp. 345–375). Academic Press. https://doi.org/10.1016/B978-0-443-16030-1.00007-9

Bhagat, S., Joshi, A., & Jain, A. (2024). Processes and mechanism involved in effective bioremediation of xenobiotic substances from agricultural fields. In *Microbiome-assisted bioremediation* (pp. 81–101). Academic Press. https://doi.org/10.1016/B978-0-443-21911-5.00017-9

Bonfante, A., Monaco, E., Manna, P., de Mascellis, R., Basile, A., Buonanno, M., Cantilena, G., Esposito, A., Tedeschi, A., de Michele, C., Belfiore, O., Catapano, I., Ludeno, G., Salinas, K., & Brook, A. (2019). LCIS DSS—An irrigation supporting system for water use efficiency improvement in precision agriculture: A maize case study. *Agricultural Systems, 176*, Article 102646. https://doi.org/10.1016/j.agsy.2019.102646

Brancalion, P. H. S., & Holl, K. D. (2020). Guidance for successful tree planting initiatives. *Journal of Applied Ecology, 57*(12), 2349–2361. https://doi.org/10.1111/1365-2664.13725

Chinseu, E., Dougill, A., & Stringer, L. (2019). Why do smallholder farmers dis-adopt conservation agriculture? Insights from Malawi. *Land Degradation & Development, 30*(5), 533–543. https://doi.org/10.1002/ldr.3190

El Rasafi, T., Pereira, R., Pinto, G., Gonçalves, F. J. M., Haddioui, A., Ksibi, M., Römbke, J., Sousa, J. P., & Marques, C. R. (2021). Potential of Eucalyptus globulus for the phytoremediation of metals in a Moroccan iron mine soil—a case study. *Environmental Science and Pollution Research, 28*(13), 15782–15793. https://doi.org/10.1007/s11356-020-11494-5

Gonzalez-Ollauri, A. (2022). Sustainable use of nature-based solutions for slope protection and erosion control. *Sustainability, 14*(4), 1981. https://doi.org/10.3390/su14041981

Graham, S., Ihli, H. J., & Gassner, A. (2022). Agroforestry, indigenous tree cover and biodiversity conservation: A case study of Mount Elgon in Uganda. *The European Journal of Development Research, 34*(4), 1893–1911. https://doi.org/10.1057/s41287-021-00446-5

Gross, A., & Glaser, B. (2021). Meta-analysis on how manure application changes soil organic carbon storage. *Scientific Reports, 11*(1), 5516. https://doi.org/10.1038/s41598-021-82739-7

Gunina, A., & Kuzyakov, Y. (2022). From energy to (soil organic) matter. *Global Change Biology, 28*(7), 2169–2182. https://doi.org/10.1111/gcb.16071

Johnson, C., Tilt, J. H., Ries, P. D., & Shindler, B. (2019). Continuing professional education for green infrastructure: Fostering collaboration through interdisciplinary trainings. *Urban Forestry & Urban Greening, 41*, 283–291. https://doi.org/10.1016/j.ufug.2019.04.021

Lelamo, L. L. (2021). A review on the indigenous multipurpose agroforestry tree species in Ethiopia: Management, their productive and service roles and constraints. *Heliyon, 7*(9), Article e07874. https://doi.org/10.1016/j.heliyon.2021.e07874

Li, S., Liang, G., Wu, X., Lu, J., Plougonven, E., Wu, H., & Liu, X. (2021). Soil water repellency influences maize yield by changing soil water availability under long-term tillage management. *Soil Discussions* [preprint]. https://doi.org/10.5194/soil-2021-57

Löbmann, M. T., Maring, L., Prokop, G., Brils, J., Bender, J., Bispo, A., & Helming, K. (2022). Systems knowledge for sustainable soil and land management. *Science of the Total Environment, 822*, Article 153389. https://doi.org/10.1016/j.scitotenv.2022.153389

Lu, C., Hong, Y., Liu, J., Gao, Y., Ma, Z., Yang, B., & Waigi, M. G. (2019). A PAH-degrading bacterial community enriched with contaminated agricultural soil and its utility for microbial bioremediation. *Environmental Pollution, 251*, 773–782, ISSN 0269-7491. https://doi.org/10.1016/j.envpol.2019.05.044

Luo, Q., Zhen, L., Xiao, Y., & Wang, H. (2020). The effects of different types of vegetation restoration on wind erosion prevention: A case study in Yanchi. *Environmental Research Letters, 15*(11), Article 115001. https://doi.org/10.1088/1748-9326/abbaff

Moyo, H., Slotow, R., Rouget, M., Mugwedi, L., Douwes, E., Tsvuura, Z., & Tshabalala, T. (2021). Adaptive management in restoration initiatives: Lessons learned from some of South Africa's projects. *South African Journal of Botany, 139*, 352–361. https://doi.org/10.1016/j.sajb.2021.03.016

Nadi, F. (2023). Presenting a new approach for energy-exergy-environmental-economic evaluation of agroecosystems: A case study of the mung bean crop rotation. *Clean Technologies and Environmental Policy, 25*(10), 3437–3450. https://doi.org/10.1007/s10098-023-02614-z

Oral, H. V., Carvalho, P., Gajewska, M., Ursino, N., Masi, F., Hullebusch, E. D. V., & Zimmermann, M. (2020). A review of nature-based solutions for urban water management in European circular cities: A critical assessment based on case studies and literature. *Blue-Green Systems, 2*(1), 112–136. https://doi.org/10.2166/bgs.2020.932

Shah, K. K., Modi, B., Pandey, H. P., Subedi, A., Aryal, G., Pandey, M., & Shrestha, J. (2021). Diversified crop rotation: An approach for sustainable agriculture production. *Advances in Agriculture, 2021*, 1–9. https://doi.org/10.1155/2021/8924087

Siyar, R., Doulati Ardejani, F., Norouzi, P., Maghsoudy, S., Yavarzadeh, M., Taherdangkoo, R., & Butscher, C. (2022). Phytoremediation potential of native hyperaccumulator plants growing on heavy metal-contaminated soil of khatunabad copper smelter and refinery, Iran. *Water, 14*(22), 3597. https://doi.org/10.3390/w14223597

Smułek, W., Sydow, M., Zabielska-Matejuk, J., & Kaczorek, E. (2020). Bacteria involved in biodegradation of creosote PAH—A case study of long-term contaminated industrial area. *Ecotoxicology and Environmental Safety, 187*, Article 109843. https://doi.org/10.1016/j.ecoenv.2019.109843

Udawatta, R., Rankoth Lalit, M., & Jose Shibu. (2021). Agroforestry for biodiversity conservation. In *Agroforestry and ecosystem services* (pp. 251–274). Springer International Publishing. https://doi.org/10.1007/978-3-030-80060-4

Willson, R. (2021). The importance of knowledge exchange as a method to help farmers transition to sustainable farming methods and mitigate climate change. https://doi.org/10.24382/373

Wood, S. A., & Bowman, M. (2021). Large-scale farmer-led experiment demonstrates positive impact of cover crops on multiple soil health indicators. *Nature Food, 2*(2), 97–103. https://doi.org/10.1038/s43016-021-00222-y

Yashmita-Ulman, Singh, M., Kumar, A., & Sharma, M. (2021). Conservation of plant diversity in agroforestry systems in a biodiversity hotspot region of Northeast India. *Agricultural Research, 10*(4), 569–581. https://doi.org/10.1007/s40003-020-00525-9

Yue, C., Jian, J., Ciais, P., et al. (2024). Field experiments show no consistent reductions in soil microbial carbon in response to warming. *Nature Communications, 15*, 1731. https://doi.org/10.1038/s41467-024-45508-4

Zhao, J., Yang, Y., Zhang, K., Jeong, J., Zeng, Z., & Zang, H. (2020). Does crop rotation yield more in China? A meta-analysis. *Field Crops Research, 245*, Article 107659. https://doi.org/10.1016/j.fcr.2019.107659

Influence of Cropping System on Soil Properties

Ahmed Abdul Haleem Khan

Abstract

Soil is one among the important components of the planet Earth along with air and water. The mankind is bestowed by soil to supply food, clothes, and shelter. The demand for increased crop yields to feed the growing world's population with limited fertile farming land the challenge has become everlasting. The majority of the world population resides in semi-arid and arid climatic environments that constitute forty percent of landmass. The land portion utilized for cropping is under hindrances of typical situations that need to be managed by up gradation of technology available in contemporary world. The use of reduced synthetic agrochemicals, increase the input of organic manures, apply microbial inoculants to improve soil fertility, and use of green manures, biochar and sustain the farming system with better yields.

Keywords

Biochar · Soil · Climate · Environment · Cropping system · Technology

1 Introduction

Soil is one among the important components of the planet Earth along with air and water. This is principal part that interacts with water and air to support micro and macro forms of living organisms (bacteria, fungi, viruses, protozoans, nematodes, plants, and animals). The mankind is bestowed by soil to supply food, clothes, and shelter. Though trapping the light energy into chemical energy by plant forms provides food and feed; most of the terrestrial plants grow on soil as substrate. The plants grow in both wild and cultivated modes; the cultivated forms of plants were grown on land close to water bodies. The gradual cultivation practices were enhanced with wide varieties of plant varieties as farms and orchards in plains. There are crop plants of different terms, i.e., short and long term with varied nutrient requirements. The nutrients could be required in trace to bulk amount that depends on the soil quality and crop need. The conventional farming practices use of tillage, harrowing, and rain feeding for water that was mechanized with progress in time. The soil provides a variety of ecosystem services that sustain life on the earth (Fig. 1).

The mechanization of agriculture gradually degraded the soil quality and green revolution brought agrochemicals that increased yields with burden of soil contamination by several pollutants. These pollutants were complex hydrocarbons that remain in soil and interfere with the structure due to the non-availability of decomposing microbes. These chemical pollutants also affect the beneficial organisms in soil that support the plants. There are multiple approaches used to increase the crop yields and save the environment from the curb of soil infertility. This chapter is intended to present the scenario around the globe with different strategies being reported for cropping practices that retain soil health in the available scientific literature from public domain.

1.1 Intercropping

The monoculture in cropping was reported for a variety of negative effects, i.e., contamination of soil and water, reduction in biodiversity, soil organic content, erosion frequency, and disease resistance by pests and phytopathogens. The farming land is used to cultivate diverse crops at a time in

A. A. H. Khan (✉)
Department of Botany, Telangana University, Nizamabad, Telangana State, India
e-mail: aahaleemkhan@gmail.com

J. A. Parray (ed.), *Soil and Land Use Change*, Sustainable Landscape Planning and Natural Resources Management,
https://doi.org/10.1007/978-3-031-93075-1_6

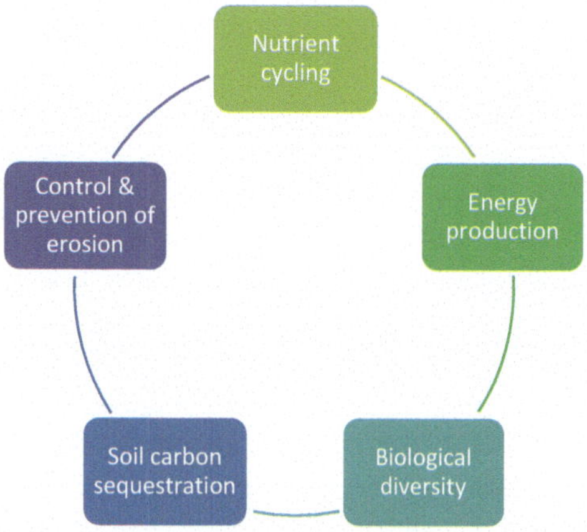

Fig. 1 Physical properties of soil with impact on ecosystem services

vicinity by selection of proper plant species is called intercropping. The cultivation of crops with diversification showed enhanced productivity, nutrient availability, and tolerance to drought and incidences of plant diseases. The intercropping with proper selection of plants was known to cut down the expenses for agrochemicals. The microbial communities of soil from diverse plant species stimulate the availability of mineral nutrients (phosphorus, potassium, iron, and calcium) due to solubilization activity.

The intercropping practice was reported for the concurrent usage of multiple crop cultivation in the same farming land area with efficient exploitation of water and soil resources. Among the variety of crop combination farming cereal-legume intercropping supports the improved yields with enhanced nitrogen supply to cereal crops from legume crop root system. Lopez-Bellido et al. (2024) evaluated the intercropping of wheat (*Triticum aestivum*)-faba bean (*Vicia faba*) on Vertisol under rainfed conditions. The findings of the study showed high grain yield of wheat than monoculture. These results found faba bean improved the soil nitrogen levels that supported wheat crop. The intercropping studies were investigated on faba bean along with cereal grain (*T. aestivum/Avena sativa*) cultivation for crop yield and weed control. The combination of beans and oats was found to show high yield and weeds were reduced during the intercropping (Boutagayout et al., 2024).

The monocrop and intercropping (mixed and row) of broccoli (*Brassica oleracea*) and fava bean (*Vicia faba*) was studied for physico chemical properties of soil. The findings showed increase in nitrogen, phosphorus, and potassium levels in intercropping practices. The intercropping of tomato-potato and onion increased beneficial microorganisms with decline in plant diseases. The capsicum and dandelion intercropping reduced disease incidence of *Phytophthora* (Li et al., 2023).

1.2 Integrated Cropping

The study reported on camelina (*Camelina sativa*), sorghum (*Sorghum bicolor*), and wheat (*Triticum aestivum*) integrated dryland (rain-fed) cropping in different crop rotations to explore the impact on soil properties (water content, organic carbon, microbial biomass carbon, and nitrogen) and crop productivity. The soil properties were increased by crop rotations with oilseed crop, camelina (Obeng et al., 2024). Ambus et al. (2018) carried investigated integrated cropping of soybean (*Glycine max*), black oat (*Avena strigosa*), rye grass (*Lolium multiflorum*), and livestock grazing for changes in soil structure. The moderate grazing by livestock was suggested to support the soil structure. The jute crop was sown in different combinations like potato, rice, rapeseed, and sesame. The findings showed better soil properties that increased yields of potato combination (Biswas et al., 2006).

1.3 Organic Cropping

The soil is a supporting medium for living organisms on the Earth. The soil health depends on a variety of functions, i.e., parent rock material, topography, and local climatic conditions that support biodiversity. The intensified cropping systems with monoculture and tillage, ploughing hampers the crop yields. Prior to invention of synthetic agrochemicals, animal manure, and domestic waste-based composts were applied as soil additives that played important role in improving the crop yields. The synthetic agrochemicals soon replaced the compost and manures that had great impact on enhanced yields. The long-term use of synthetic components increased the negative impact on the soil ecosystem. To overcome the burden of synthetic farm chemicals, organic farming was brought to limelight with use of farmyard manure. The organic farming was reported to improve soil pH, reduce bulk density, and enhance water-holding capacity, and microbial biomass with increase in soil carbon and nitrogen content.

Bulluck et al. (2002) showed use of organic farming improved soil microbial activity that reduced the plant pathogens, i.e., *Pythium* and *Phytophthora* sp. due to antagonist *Trcihoderma* sp. and thermophilic microbes. The combined cropping of soybean—mustard with use of organic (farmyard) manure improved soil properties. The olive orchards, cashews, and potato fields under organic farming resulted in better yields with enhanced soil enzyme activities, microbial biomass, and mineral content. The findings of Lu et al. (2020) showed increase in species diversity in organic farms compared to conventional farms. The combined organic farming of tomato (*Solanum lycopersicum*) and carrot (*Daucus carota*) showed high yields than control. The cropping of rice and wheat by organic approach was studied for soil quality and yields of crops. The practice enhanced soil organic matter and crop productivity through organic farming compared to conventional farming (Sahu et al., 2024).

1.4 Crop Rotation

The cultivation of crops for long term hampers the soil fertility and reduces the yield with impact on soil microbial communities. The continuous cropping of sugar beet decreased disease resistance by plant pests with drastic change in the yields (Huang et al., 2021). The study carried on rubber dandelion (*Taraxacum kok-saghyz*) and sugar beet (*Beta vulgaris*) on crop rotation increased soil organic matter, minerals, and microbial communities (Guo et al., 2024a, 2024b). The five-year term crop rotations were carried to assess soil health indicators. The important key soil health indicators were identified. The study on crop rotation from paddy to maize and its impact on carbon release was carried in greenhouse using hydroponic culture. The results showed rotation of crop developed mineral deposits (He et al., 2015).

1.5 Application of Manures

The irregularities in nutrients in soil result in reduced buffering action and decreased nutrient holding tendency that reduces microbial activities (Fig. 2). There are variety of manures, i.e., press mud, poultry refuse, vermicompost, and farmyard manure that proved far safer than chemical fertilizers (Borase et al., 2020). The manure application in soil of farming area was reported to increase organic matter in perennial grasses and reduced by cereal-legume cropping. The sandy and eroded soils applied with manure for long term were reported for increase in water retention capacity and hydraulic properties of soil. The crops like wheat, barley, maize, and clover were studied for manure application and its influence on soil properties. The manure improved physicochemical properties of soil (silt clay, silt loam, and sandy loam) (Yan et al., 2023).

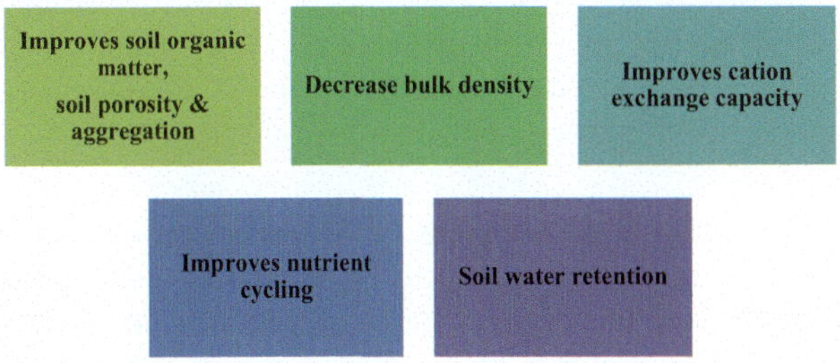

Fig. 2 Influences of manure application to crop soil

The pearl millet-wheat was cropped with application of organic manures (poultry manure, farmyard manure, and press mud) to evaluate the soil organic carbon and properties of microbial communities. The results showed farmyard manure improved soil properties and microbial biomass (Kumari et al., 2024). The study in greenhouse on cultivation vegetables as crop rotation to rice–wheat was evaluated for soil properties after use of organic fertilizers. The results show application of organic fertilizers controls the degradation of soil quality. The garden pea-french bean was cultivated using farmyard manure and evaluated for crop yield. The recommended dose of farmyard manure that improved the soil properties and yield of crops was found in this study (Mahanta et al., 2013).

1.6 Cover Crops

Conventional farming for long term brings the degradation of soil physicochemical properties. There urge to reduce the burden of soil damage by routine practices could be done by cover cropping. Cover cropping is an important practice to avoid leaching nutrients, soil erosion, enhance sequestration of carbon, and increase the microbial biomass, soil enzyme activities and suppression of weeds by allelopathic effects. The study evaluated the effect of different cover crops (*Raphanus sativus, Fagopyrum esculentum, Linum usitatissimum, Guizotia abyssinica,* and *Avena strigosa*) on allelopathic effects on weeds and the findings signify the cover crop strongly prevent the weed population (Sturm et al., 2018).

The impact of diverse cropping wheat (*Triticum aestivum*) followed by winter pea (*Pisum sativum*) was investigated for soil properties and microbial communities. The findings of the study showed improved soil health properties due to the cover crop, i.e., pea (Domnariu et al., 2024). The different cover crops were sown in dryland, i.e., radish (*Raphanus sativus*) hairy vetch (*Vicia villosa*) as monoculture and mix culture of two crops to evaluate the impact on soil microbial communities and multiple functions of soil. The study suggested cover crop enhanced richness of beneficial microbes in soil with mix culture of radish and hairy vetch (Gui-long et al., 2024). The study was carried on cereal rye (*Secale cereale*), camelina (*Camelina sativa*) as cover crop introduced in place of corn (*Zea mays*) and soybean (*Glycine max*) to evaluate the effects on root disease (*Pythium*). The findings of study showed cover crop of camelina reduced the disease incidence in corn (Acharya et al., 2020).

The cash crops, i.e., asparagus, carrot, onion, and sugar beet were evaluated by use of seven cover crop species (niger seed, phacelia, rapeseed, radish, vetch, black oat, and buckwheat) to avoid the root-borne disease incidences. The cover crop plants strongly controlled the root-borne fungal pathogenic fungi. The study evaluated the use of microbial inoculants to decompose the cover crop residues and the findings showed no significant effects of inoculants were detected (Oberholzer et al., 2024). The cover crops, i.e., millet (*Pennisetum glaucum*), sun hemp *(Crotalaria ochroleuca*) pigeon pea (*Cajanus cajans*), and perennial grass (*Urochola ruziziensis*) were rotated with rice (*Oryza sativa*) and soybean (*Glycine max*) for its impact on chemical and physical properties of soil. The cover crops improved the yields of soybean and rice due to better soil properties (Nascente & Stone, 2018).

1.7 Intensified Cropping

The intensified approach of cropping permits to cultivate multiple crop plants in limited piece of farmland and gains more benefits. The simultaneous growing of cereals and legume crops proved better yields than chemical-based agriculture. The crops with short duration were reported to conserve land and water in farming soils. The study carried out by Yadav et al. (2023) on maize (*Zea mays*) in combination with sweet potato (*Ipomoea batatas*), colocasia (*Colocasia esculenta*), turmeric (*Curcuma longa*), ginger (*Zingiber officinale*) for biomass recycling and resource conservation. The crop intensification showed the combination of maize-sweet potato improved the soil health and better crop yields. The dryland agriculture was evaluated with intensification of crop that showed increased residue biomass and water capture in soil (Shaver et al., 2003).

The main crop, i.e., maize (*Zea mays*) is sown in different combinations–cowpea (*Vigna unguiculata*), rapeseed (*Brassica campestris*), buckwheat (*Fagopyrum esculentum*), barley (*Hordeum vulgare*), garden pea (*Pisum sativum*) and rajmash (*Phaseolus vulgaris*) for soil moisture and improve the soil health. The combination of cowpea with maize improved the soil health and increased the crop yields (Singh et al., 2021a, 2021b). The intensification of wheat (*Triticum aestivum*) with different legumes, i.e., chickpea (*Cicer arietinum*), lentil (*Lens culinaris*), pea (*Pisum sativum*) was assessed for better yields. The legume crops along with wheat showed improved soil properties and enhanced crop yields.

1.8 Agroforestry

This approach is use of land for cultivation of different types of perennial plant species along the short-term crop plants. This type of plantation provides interaction between woody and non-woody plant species. This system of cultivation gained wide popularity around globe as it supports sustainable farming and restoration of ecosystem by prevention of soil erosion, maintaining soil moisture, control soil pH, and enhance physical and chemical properties of soil. The study on different types of agroforestry, i.e., home garden, woodlot, and parkland when evaluated for soil properties and the findings showed home garden kind of agroforestry increased accumulation of nitrogen with boost in soil health (Madalcho & Tefera, 2015).

The samples of soil from agroforestry zones planted with trees, i.e., *Cordia africana, Ficus sur,* and *Manihot esculenta* were evaluated for chemical properties. The interactions between the plants improved pH, ion exchange, and better soil health (Gota et al., 2024). The agroforestry system with plantation of banana and grevillea (*Grevillea robusta*) were examined for soil properties. The intercropping showed better soil health than sole cropping of target plants (Musongora et al., 2023). The agroforestry system with *Grevillea robusta*, banana (*Musa* spp.), and beans (*Phaseolus vulgaris*) were studied for agronomic and soil properties. The soil fertility was increased by cultivation of beans along with other plants. The agroforestry systems with *Grevillea robusta* were examined by growing maize for its productivity. The shade in this system was not beneficial to maize yields (Lott et al., 2009).

1.9 Impact of Microbes on Soil

The soil quality is improved by several methods such as crop rotation, green (plant based) fertilizers, no or reduced tillage, and cultivation of cover crops. The soil quality is quantified by different strategies, among them biological approach (enzyme activity) is prominent. The soil quality is commonly assessed by measurement of chemical, physical and biological factors. The biological quality of soil is usually represented by enzyme activities (Liu et al., 2023). The soil enzymes such as β glucosidase and dehydrogenase are more sensitive and influence the soil organic matter. The nitrogen content and enzyme assays were reported to assess the microbial activity in soil. The microbial communities of soil are known to contribute ecosystem services via biogeochemical cycles of minerals. The microbial activities in the soil improve nutrients and suppress disease pathogens that result in decline of soil-borne diseases.

The impact of intercropping of melon (*Cucumis melo*)—Cowpea (*Vigna unguiculata*) in organic farming improved the mineral content and beneficial microbial communities than monocrop of the crops (Cuartero et al., 2022). The study compared combined cropping with organic farming of vegetable (onion-lettuce, lettuce-leek) production with set of conditions like application of plant-based fertilizers, animal manure, and reduced tillage in sandy loam soil for assessing soil microbial activity. The findings showed beneficial effects of cropping in terms of improved soil microbial activity and soil mineral content with better yields of crops (Hefner et al., 2024).

The symbioses of fungi with root zone of growing plants were reported for increase in phosphorus and water-holding capacity of soil. The study evaluated growth and development of Indian pennywort (*Centella asiatica*) in combination with arbuscular mycorrhizae (*Acaulospora, Funneliformis, Gigaspora, Scutellospora* sp.) under water deficit. The colonization of inoculated fungi improved the yields in water deficit stress (Praseartkul et al., 2024). The intercropping of three different crops (*Crocus sativus, Vicia sativa, and Lavandula intermedia*) were grown in olive groves fed by rainwater to compare with conventional cultivation practices. The alley cropping resulted in abundances of soil microbial communities that promoted plant growth with increased soil mineral content (Aguilera-Huertas et al., 2023).

The crop plants, i.e., soybean and corn were grown in plots that were treated with fertilizers mixed with municipal biosolids to explore the availability of soil bacterial communities. The biosolids applied enhanced the beneficial bacterial communities and supported to replace the chemical fertilizers. The monoculture and mixed cropping of barley (*Hordeum vulgare*) and pea (*Pisum sativum*) with added fertilizers for investigation of bacteria and fungi colonized in the soil. The findings showed increase in bacterial levels were high in mixed cropping (Guo et al., 2024a, 2024b). The four different locations of maize (*Zea mays*) cultivation were studied for soil-associated prokaryotic microbes and properties of soil. The parameters such pH of soil, cropping approach, and location of crop sown were reported to be important in the colonization of prokaryotes in that soil (Xue-liang et al., 2022). The monocultures of macadamia (*Macadamia integrifolia*) and intercropping with konjac (*Amorphophallus konjac*), dasheen (*Colocasia esculenta*), and maize (*Zea mays*) were evaluated for soil organic carbon and microbial communities in agroforestry setup. The soil fungal communities were predominant to contribute the organic carbon content to soil.

1.9.1 Tillage and Soil Properties

The tillage is agronomic practice that loosens the soil layers to form complex ecosystem for seedbed making. The technique of tillage involves use of harrow, plough, and pulverizing device to loosen the soil crust. The prolonged tillage practice was reported to show detrimental effects on soil quality and structure. The different tillage strategies, i.e., conventional, reduced and no tillage were assessed in three cropping systems. The crops in this study were wheat, lentil, barley, oat, and pea sown with types of tillage to explore the yield and soil properties. The reduced and no tillage increased soil nutrient levels (magnesium, sodium, and phosphorus) and moisture content.

The crop productivity and physicochemical properties of soil were evaluated by cultivation of pigeon pea (*Cajanus cajan*) under broad bed furrow, minimum, conventional and zero tillage. The broad bed furrow approach was shown beneficial with high crop yield and improved soil properties (Singh et al., 2018). The report on different types of crops, i.e., sorghum, cowpea, soybean as monocrop and intercrop with different tillage practices to study the impact on physical and chemical properties of soil by Akinola et al. (2023). The results suggested that diverse crop cultivation preserves the soil properties. Alam et al. (2024) studied mustard-mung bean cultivation with different tillage approaches, i.e., minimum, conventional, and deep tillage. Among the tillage methods applied, minimum tillage was satisfactory in soil improvement with increased crop yield. The effects of tillage on crop yield and soil fertility in monoculture of rice crop was investigated. The study reports minimum tillage provides better yield than conventional approach (Vitali et al., 2024). Zhang et al. (2015) quantified the role of multiple cropping systems (rice–wheat) with long-term tillage for greenhouse gas emissions. The rotary tillage practice proved beneficial to reduce the gas emissions and improve the crop productivity.

1.9.2 Irrigation with Saline Water

The gradual decrease in fresh water for irrigation of crop plants resulted from surface and ground water scarcity. The saline water used in place of fresh water in farming accumulated salt ions with altered physical and chemical properties of soil. The use of drip irrigation with saline water was reported to affect the soil environment by negative changes in soil properties. The study by Singh et al., (2021a, 2021b) reported enhanced salt ions that affected soil microbial activity. The furrow irrigation with saline water in maize resulted in decrease in crop yield and increase in soil saline levels. The study on use of saline water with different electrical conductivities was employed to irrigate the maize and wheat crop irrigation. The soil quality parameters showed safe electrical conductivity limits for saline water irrigation (Wang et al., 2023).

1.9.3 Biochar

The biomass treated at temperature more than 250° C in low oxygen availability produces carbon-rich material termed as biochar. This material is reported to improve the soil quality that shows enhanced crop productivity (Fig. 3). The soil enzymes involved in the biogeochemical cycles were increased in presence of biochar.

Baldoni et al. (2023) studied the effect of different doses of wood gasification biochar applied in soil with wheat cultivation for assessment of soil properties. The treated and

Fig. 3 Properties of biochar with beneficial and negative effects in soil

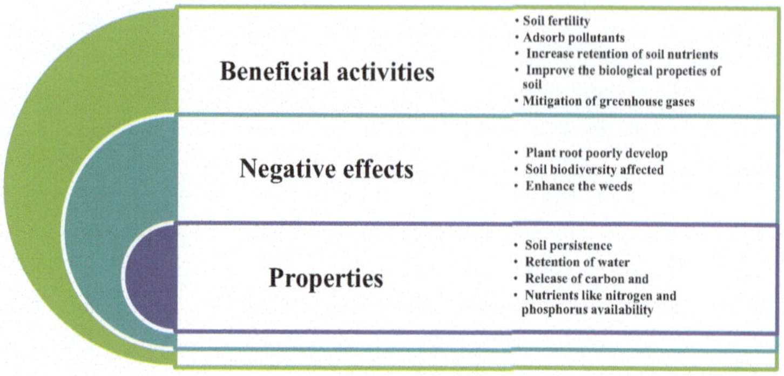

untreated soils were found to exhibit similarities in soil properties.

1.9.4 Denitrification Inhibitors

The nitrogen content in soil is important parameter for its fertility and supports more plant life. The soil contains nitrates and nitrites as nitrogen components that supply to plants. The biogeochemical cycle, i.e., nitrogen cycle is crucial in maintaining the nitrogen content in soil. The microbial conversions change the fixed form to free state of nitrogen (N_2) by denitrification. This is anaerobic process that causes the loss of nitrogen from soil. The plant roots release allelochemicals that prevent the loss of nitrogen through denitrification. The roots of *Fallopia* sp. showed release of secondary metabolites that were inhibitory against the denitrifying bacteria in soil. The application of procyanidin to paddy soil was reported to reduce the emission of nitrous oxide (N_2O) by inhibiting the activity of denitrifying bacteria. The different nitrogen-supplying fertilizers applied with procyanidin in fields of wheat and rice in winter and summer seasons to check the ability to avoid the denitrification. The results showed effective role of procyanidin in prevention of nitrogen loss from soil (Ye et al., 2022). The exogenous application of procyanidin in field of lettuce (*Lactuca sativa*), and found the increase in nitrate levels around the roots than control plants (Galland et al., 2019). The studies under aerobic and anaerobic conditions in corn-cultivated soils were treated with grape seed extracts and procyanidin for measurement of nitrous oxide and inhibition of denitrification. The grape seed extract proved to be inhibitor of nitrous oxide emissions and was found as soil amendment with applied value (Hsiao et al., 2024).

2 Conclusion

The demand for increased crop yields to feed the growing world's population with limited fertile farming land the challenge has become everlasting. The scientific community is chasing this challenge with variety of solutions, i.e., intercropping, integrated cropping, cover crops, crop rotation, agroforestry, and use of greener approaches to gain more benefits in cropping system. The majority of world's population resides in semi-arid and arid climatic environments that constitute forty percent of land mass. The land portion utilized for cropping is under hindrances of typical situations that need to be managed by upgradation of technology available in contemporary world. The use of reduced synthetic agrochemicals increase the input of organic manures, apply microbial inoculants to improve soil fertility, and use of green manures, biochar and sustain the farming system with better yields.

References

Acharya, J., Moorman, T. B., Kaspar, T. C., Lenssen, A. W., & Robertson, A. E. (2020). Cover crop rotation effects on growth and development, seedling disease, and yield of corn and soybean. *Plant Disease, 104*, 677–687.

Aguilera-Huertas, J., Cuartero, J., Ros, M., Pascual, J. A., Parras-Alcantara, L., Gonzalez-Rosado, M., Ozbolat, O., Zornoza, R., Ege-Cortines, M., Hurtado-Navarro, M., & Lozano-Garcia, B. (2023). How binomial (traditional rainfed olive grove-*Crocus sativus*) crops impact the soil bacterial community and enhance microbial capacities. *Journal of Environmental Management, 345*, Article 118572.

Akinola, F. F., Fasinmirin, J. T., Olanrewaju, O. O., Ewulu, B. S., & Olorunfemi, I. E. (2023). Changes in soil physical and mechanical properties under different tillage and cropping systems in alfisol soil on southwestern Nigeria. *Farming Systems, 1*, Article 100050.

Alam, M. J., Islam, M. S., Mondol, A. T. M. A. I., Naser, H. M., Salahin, N., Alam, M. K., Islam, M. M., Akter, S., & Alam, Z. (2024). Cropping system-based fertilizer strategies for crop productivity and soil health under minimum tillage in grey terrace soil. *Heliyon, 10*, Article e24106.

Ambus, J. V., Reichert, J. M., Gubiani, P. I., & Carvalho, P. C. F. (2018). Changes in composition and functional soil properties in long-term no-till integrated crop-livestock system. *Geoderma, 330*, 232–243.

Baldoni, N., Francioni, M., Trozzo, L., Toderi, M., Fornasier, F., D'Ottavio, P., Corti, G., & Cocco, S. (2023). Effect of wood gasification biochar on soil physicochemical properties and enzyme activities, and on crop yield in a wheat-production system with sub-alkaline soil. *Biomass and Bioenergy, 176*, Article 106914.

Biswas, B., Ghosh, D. C., Dasgupta, M. K., Trivedi, N., Timsina, J., & Dobermann, A. (2006). Integrated assessment of cropping systems in the Eastern Indo-Gangetic plain. *Field Crops Research, 99*(1), 35–47.

Borase, D. N., Nath, C. P., Hazra, K. K., Senthil, M., Singh, S. S., Praharaj, C. S., Singh, U., & Kumar, N. (2020). Long-term impact of diversified crop rotations and nutrient management practices on soil microbial functions and soil enzymes activity. *Ecological Indicators, 114*, Article 106322.

Boutagayout, A., Belmalha, S., Nassiri, L., El Alami, N., Jiang, Y., Lahlali, R., & Bouiamrine, E. H. (2024). Weed competition, land equivalent ratio and yield potential of faba bean (*Vicia faba* L.)—cereals (*Triticum aestivum* L. and/or *Avena sativa* L.) intercropping under low-input conditions in Meknes region, Morocco. *Vegetos, 37*, 866–879.

Bulluck, L. R., III., Brosius, M., Evanylo, G. K., & Ristaino, J. B. (2002). Organic and synthetic fertility amendments influence soil microbial, physical and chemical properties on organic and conventional farms. *Applied Soil Ecology, 19*(2), 147–160.

Cuartero, J., Pascual, J. A., Vivo, J. M., Ozbolat, O., Sanchez-Navarro, V., Egea-Cortines, M., Zornoza, R., Martinez-Mena, M., Garcia, E., & Ros, M. (2022). A first-year melon/cowpea intercropping system improves soil nutrients and changes the soil microbial community. *Agriculture, Ecosystems & Environment, 328*, Article 107856.

Domnariu, H., Reardon, C. L., Manning, V. A., Gollany, H. T., & Trippe, K. M. (2024). Legume cover cropping and nitrogen fertilization influence soil prokaryotes and increase carbon content in dryland wheat systems. *Agriculture, Ecosystems & Environment, 367*, Article 108959.

Galland, W., Piola, F., Burlet, A., Mathieu, C., Nardy, M., Poussineau, S., & el Zahar Haichar, F. (2019). Biological denitrification inhibition (BDI) in the field: A strategy to improve plant nutrition and growth. *Soil Biology & Biochemistry, 136*, 107513.

Gota, H. G., Madalcho, A. B., Kerse, B. L., Szwagrzyk, J., & Solomon, T. (2024). The impact of native trees, *Cordia africana* and *Ficus sur*, and the economically valuable *Manihot esculenta* on soil chemical properties in an agroforestry system. *Trees, Forests and People, 15*, Article 100471.

Gui-long, L., Xiao-fen, C., Wen-jing, Q., Jing-rui, C., Ke, L., Lu-yuan, S., Ming, L., Meng, W., Jian-bo, F., Chang-xu, X., & Jia, L. (2024). Characteristics of the microbial communities regulate soil multifunctionality under different cover crop amendments in ultisol. *Journal of Integrative Agriculture, 23*(6), 2099–2111.

Guo, C., Yang, C., Fu, J., Song, Y., Chen, S., Li, H., & Ma, C. (2024a). Effects of crop rotation on sugar beet growth through improving soil physicochemical properties and microbiome. *Industrial Crops & Products, 212*, Article 118331.

Guo, T., Yao, X., Wu, K., Guo, A., & Yao, Y. (2024b). Response of the rhizosphere soil microbial diversity to different nitrogen and phosphorus application rates in a hulless barley and pea mixed-cropping system. *Applied Soil Ecology, 195*, Article 105262.

He, Y., Siemens, J., Amelung, W., Goldbach, H., Wassmann, R., Alberto, M. C. R., Lucke, A., & Lehndorff, E. (2015). Carbon release from rice roots under paddy rice and maize-paddy rice cropping. *Agriculture, Ecosystems & Environment, 210*, 15–24.

Hefner, M., Gebremikael, M. T., & Kristensen, H. L. (2024). Soil microbial activity improved while intensifying vegetable production by use of plant based fertilisers, cover crops and reduced tillage. *Pedobiologia–Journal of Soil Ecology, 102*, 150926.

Hsiao, C. J., Frie, A., Mitchell, S., Venterea, R., & Griffis, T. (2024). Efficacy of grape seed procyanidins for inhibiting denitrification varies by source, soil texture, and cropping history. *Applied Soil Biology, 195*, Article 105254.

Huang, W., Sun, D., Wang, R., & An, Y. (2021). Integration of transcriptomics and metabolomics reveals the responses of sugar beet to continuous cropping obstacle. *Frontiers in Plant Science, 12*, Article 711333.

Kumari, M., Sheoran, S., Prakash, D., Yadav, D. B., Yadav, P. K., Jat, M. K., & Ankit, A. (2024). Long-term application of organic manures and chemical fertilizers improve the organic carbon and microbiological properties of soil under pearl millet-wheat cropping system in North-Western India. *Heliyon, 10*, Article e25333.

Li, P., Liu, H., Chen, Y., Zhang, X., Cao, N., Sun, Y., Jia, M., Wu, M., Tong, X., Jiang, X., Yu, X., & Cheng, Y. (2023). Effect of dandelion (*Taraxacum mongolicum* Hand-Mazz.) intercropping with different plant spacing on blight and growth of pepper (*Capsicum annuum* L.). *Phyton-International Journal of Experimental Botany, 92*(8), 2227–2244.

Liu, Q., Zhao, Y., Li, T., Chen, L., Chen, Y., & Sui, P. (2023). Changes in soil microbial biomass, diversity, and activity with crop rotation in cropping systems: A global synthesis. *Applied Soil Ecology, 186*, Article 104815.

Lopez-Bellido, R. J., Munoz-Romero, V., Fernandez-Garcia, P., & Lopez-Bellido, L. (2024). Response of wheat and faba bean to intercropping and tillage system on a Mediterranean rainfed Vertisol. *Agricultural Research, 13*, 553–562.

Lott, J. E., Ong, C. K., & Black, C. R. (2009). Understorey microclimate and crop performance in a *Grevillea robusta*-based agroforestry system in semi-arid Kenya. *Agricultural and Forest Meteorology, 149*(6–7), 1140–1151.

Lu, H. L., Chang, Y. H., & Wu, B. Y. (2020). The compare organic farm and conventional farm to improve sustainable agriculture, ecosystems, and environment. *Organic Agriculture, 10*(4), 409–418.

Madalcho, A. B., & Tefera, M. T. (2015). The effect of agroforestry practices and elevation gradients on soil chemical properties in Gununo Watershed, Ethiopia. *International Journal of Environmental and Agriculture Research, 2*, 1–7.

Mahanta, D., Bhattacharyya, R., Gopinath, K. A., Tuti, M. D., Jeevanandan, K., Chandrashekara, C., Arunkumar, R., Mina, B. L., Pandey, B. M., Mishra, P. K., Bisht, J. K., Srivastva, A. K., & Bhatt, J. C. (2013). Influence of farmyard manure application and mineral fertilization on yield sustainability, carbon sequestration potential and soil property of garden pea –french bean cropping system in the Indian Himalayas. *Scientia Horticulturae, 164*, 414–427.

Musongora, M., Karanja, N., Kimenju, W., & Kamau, S. (2023). Spatio-temporal change of selected soil physico-chemical properties in grevillea-banana agroforestry systems. *Heliyon, 9*(5), Article e16121.

Nascente, A. S., & Stone, L. F. (2018). Cover crops as affecting soil chemical and physical properties and development of upland rice and soybean cultivated in rotation. *Rice Science, 25*(6), 340–349.

Obeng, E., Obour, A. K., Nelson, N. O., Ciampitti, I. A., & Wang, D. (2024). Cropping sequence influenced crop yield, soil water, and soil properties in wheat-camelina cropping system. *Farming System, 2*, Article 100066.

Oberholzer, S., Herrmann, C., Bodenhausen, N., Krause, H. M., Mestrot, A., Speranza, C. I., & Jarosch, K. A. (2024). No effect on biological or chemical soil properties when amended with effective microorganisms for improved cover crop decomposition. *Applied Soil Ecology, 197*, Article 105358.

Praseartkul, P., Tisarum, R., Sotesaritkul, T., Chungloo, D., Theerawitaya, C., Taota, K., Singh, H. P., & Cha-um, S. (2024). Arbuscular mycorrhizal fungi improve tolerance to water deficit in Indian pennywort (*Centella asiatica*) by promoting physio-morphological and biochemical adaptations. *Agricultural Research, 13*, 718–732.

Sahu, H., Kumar, U., Mariappan, S., & Mishra, A. P. (2024). Impact of organic and inorganic farming on soil quality and crop productivity for agricultural fields: A comparative assessment. *Environmental Challenges, 15*, Article 100903.

Shaver, T. M., Petersen, G. A., & Sherrod, L. A. (2003). Cropping intensification in dryland systems improves soil physical properties: Regression relations. *Geoderma, 116*(1–2), 149–164.

Singh, Y. P., Singh, S., Nanda, P., & Singh, A. K. (2018). Impact of establishment techniques and maturity duration of pigeon pea cultivars on yield, water productivity and properties of soil. *Agricultural Research, 7*(3), 271–279.

Singh, G., Mavi, M. S., Choudhary, O. P., Gupta, N., & Singh, Y. (2021a). Rice straw biochar application to soil irrigated with saline water in a cotton-wheat system improves crop performance and soil functionality in north-west India. *Journal of Environmental Management, 295*, Article 113277.

Singh, R., Babu, S., Avasthe, R. K., Yadav, G. S., Das, A., Mohapatra, A. P., Kumar, A., Singh, V. K., & Chandra, P. (2021b). Crop productivity, soil health, and energy dynamics of Indian Himalayan intensified organic maize-based systems. *International Soil and Water Conservation Research, 9*(2), 260–270.

Sturm, D. J., Peteinatos, G., & Gerhards, R. (2018). Contribution of allelopathic effects to the overall weed suppression by different cover crops. *Weed Research, 58*, 331–337.

Vitali, A., Moretti, B., Lerda, C., Said-Pullicino, D., Celi, L., Romani, M., Fogliatto, S., & Vidotto, F. (2024). Conservation tillage in temperate in temperate rice cropping systems: Crop production and soil fertility. *Field Crops Research, 308*, Article 109276.

Wang, H., Zheng, C., Ning, S., Cao, C., Li, K., Dang, H., Wu, Y., & Zhang, J. (2023). Impacts of long-term saline water irrigation on soil properties and crop yields under maize-wheat crop rotation. *Agricultural Water Management, 286*, Article 108383.

Xue-liang, T., Jia-jia, L., Quan-cheng, L., Xin-yao, X., Yong, P., Huerta, A. I., Jian-bing, Y., Hui, L., & Wen-de, L. (2022). The effects of soil properties, cropping systems and geographic location on soil prokaryotic communities in four maize production regions across China. *Journal of Interactive Agriculture, 21*(7), 2145–2157.

Yadav, D., Singh, D., Babu, S., Madegowda, M., Singh, D., Mandal, D., Rathore, A. C., Sharma, V. K., Singhal, V., Kumawat, A., Yadav, D. K., Yadav, R. K., & Kumar, S. (2023). Intensified cropping reduces soil erosion and improves rainfall partitioning and soil properties in the marginal land of the Indian Himalayas. *International Soil and Water Conservation Research, 12*(3), 521–533.

Yan, F., Fu, Y., Paradelo, M., Zhang, F., & Arthur, E. (2023). Long-term manure and cropping systems effect on soil water vapour sorption characteristics is controlled by soil texture. *Geoderma, 436*, Article 116533.

Ye, M., Zheng, W., Yin, C., Fan, X., Chen, H., Gao, Z., Zhao, Y., & Liang, Y. (2022). The inhibitory efficacy of procyanidin on soil denitrification varies with N fertilizer type applied. *Science of the Total Environment, 806*, Article 150588.

Zhang, L., Zheng, J., Chen, L., Shen, M., Zhang, X., Zhang, M., Bain, X., Zhang, J., & Zhang, W. (2015). Integrative effects of soil tillage and straw management on crop yields and greenhouse gas emissions in a rice-wheat cropping system. *European Journal of Agronomy, 63*, 47–54.

Climate as a Driver of Biodiversity Change on Land

Menakshi Dhar, Raman Jasrotia, and Seema Langer

Abstract

Climate change poses a severe threat to global biodiversity as it has a huge impact on all levels of biological diversity from the genetic system to the whole ecosystem. The multiple aspects of climate change including increasing temperature, change in rainfall patterns, and emission in increase in emission of greenhouse gases have a great influence on all the pillars of biodiversity. According to the data given by the Intergovernmental Panel on Climate Change (2007); the global average temperature has surged by 0.2 °C per decade since the 1970s and global average precipitation has increased by 2% over the past century. Climate change alters the vegetation communities, which can affect the integrity of whole biomes. The Millennium Ecosystem Assessment projected that 5–20% of the terrestrial ecosystems on the earth may experience shifts due to these changes. The effects of climate change are projected to increase with the increase in the magnitude of climate change. Changing climatic conditions increases the potential for pests and pathogens to establish and spread and certain species that show slow reproduction or low genetic diversity are not capable enough to stay aligned with climatic change. Changes have been observed in the composition of microbial, animal, and plant communities reflecting their responses to increasing temperature. A projected change in extreme weather events has a great impact on biodiversity and ecosystems. Montane habitats and wetlands are very sensitive to climate change. Life cycle events of various floral and formal species have been altered due to changes in the onset of seasons, which make them more prone to extinction. Climate change poses an immense complication for conservation biologists in designing and executing efficient conservation strategies. The expected decline in biodiversity due to climatic shifts undermines the provision of services and the functioning of the ecosystem, potentially leading to planetary environmental upheaval. Suitable and imperative measures are required to be considered across diverse scenarios of climate disruption effects on a wide range of flora and fauna species. Primarily, an extensive multi-sectoral strategy on biodiversity issues, embracing expensive policies, stringent tactics, and programs at local, national, and international levels is crucial to address the challenges posed by the impact of climate alterations on biodiversity.

Keywords

Anthropogenic · Climate · Conservation · Flora · Fauna · Variability

1 Introduction

The average weather condition in a particular region, specified by long-term statistical patterns of various meteorological elements is referred to as the climate (WMO, 2006). However, climate change is defined as a change in the area's climate due to natural and human-induced activities. Climate change has both direct and indirect effects on the composition of species and ecosystems. Changing climatic conditions can lead to a decline in biodiversity by the mid-twenty-first century. Climate change can pose a serious threat to the various services provided by the ecosystem. The sudden alterations in the environment due to climatic changes are responsible for the destruction of the natural habitat of different floral and formal species will stop the increase in

M. Dhar · R. Jasrotia (✉) · S. Langer
Department of Zoology, University of Jammu, Jammu, Jammu and Kashmir, India
e-mail: drraman.jasrotia@jammuuniversity.ac.in

R. Jasrotia
Department of Zoology, School of Sciences, Cluster University of Jammu, Jammu, India

J. A. Parray (ed.), *Soil and Land Use Change*, Sustainable Landscape Planning and Natural Resources Management,
https://doi.org/10.1007/978-3-031-93075-1_7

temperature, alterations in rainfall patterns, ocean acidification, and numerous other extreme weather events put a huge pressure on the species which are already under the threat by anthropogenic activities. Climate change includes the long-term alterations in the weather and temperature due to human activities. Increase in temperature across the globe and extreme weather conditions are the most frequent signs of climate change. During the last three years climate change has become a global emergency, having a great impact on the well-being and sustainability of various life forms since years, the relationship between environmental conditions and numerous life forms has been in a delicate balance and was conducive for the existence of different species on the planet earth however with the beginning of an industrial revolution in the year 1850 this balance made a gradual shift and this change and changes are became evident by the mid of twentieth century. Nowadays this shift has become a cause of concern for the well-being and sustainability of different components of biodiversity (Shivanna, 2022). At present, climate change induced by different human activities has been unexampled in the last 2000 years and is increasing in all regions of the world as per the reports given by AR6 Climate Change, 2021. Climate change across the globe has a catalytic effect on environmental parameters. Biodiversity loss is responsible for food insecurity, degrades the grassland ecosystem, and causes soil and water pollution. These days we live in a digital world where climate change plays a significant role in the events occurring in any one country and has a massive effect on geographically distant regions which is evident from the crisis of the COVID-19 pandemic (Sarkar et al., 2021).

Anthropogenic activities are responsible for climatic alterations. Besides the industrial revolution, various anthropogenic activities such as intensification in agricultural operations involving the burning of fossil fuels, agricultural residue, deforestation, use of fuel-based mechanization, and the transportation sector play a critical role in changing climatic conditions. These human-induced activities are responsible for damaging global and local infrastructure, climatic catastrophes, and a decline in the total productivity of ecosystems. The increase in the consumption of energy has raised the levels of greenhouse gases that have led to atmospheric warming because much of the energy is produced due to the burning of fossil fuels in developing nations. Rising temperature has a great impact on the interior parts of any continent (Schuurmans, 2021). Changes in weather patterns due to a decline in the availability of water, melting of glaciers, and increasing temperature are responsible for causing the extinction of different plant species (Mihiretu et al., 2021). The coastal ecosystems are teetering on the brink of devastation. The temperature rise, outbreak of various insect diseases, and health-related diseases are persistent and there is a strong likelihood that a similar pattern will follow in the future due to changing climatic conditions (Hussain et al., 2018). As per the data given by Huang et al., 2022, by the year 2050, an increase in the temperature by 2 to 3% and a shift in the pattern of rainfall may have serious negative effects. Any kind of alteration in the climatic condition is responsible for the exaggeration of various pandemics and epidemics. Climate change is a global problem that involves coordination at intergovernmental level, as it influences different aspects of ecology, environment, socio-economic and socio-political disciplines.

Climate is the main environmental parameter of any area from the country and the whole world. Any kind of alteration in environmental parameters of a particular area over a period of time is responsible for climate change. These alterations have a great influence over the animals, water cycle, pattern of wind, distribution of rainfall, agricultural sector, plant growth and development and even a great negative impact on the whole food chain/food web as green plants are the food producers on the Earth's crust. The main sources of climate change are the emission of greenhouse gases from various sources such as industrialization, transportation, and urbanization that are responsible for increasing the atmospheric temperature. Greenhouse gases such as methane, carbon dioxide, hydrofluorocarbons, perfluorinated compounds, sulphur hexafluoride, and nitrous oxide have a huge impact on the environment. Among these, carbon dioxide plays a major role in the global climate change. The concentration of carbon dioxide in the troposphere varies from place to place and from time to time. Sources of carbon dioxide production are both man-made and natural. The concentration of carbon dioxide has increased by 25% over the past 125 years (Thurman and Elizabeth, 2011). This is a real and urgent challenge that is affecting the biotic components and environment across the globe. During pleasant times, climate change is a key challenge at requires to be tackled. It has increased the burden of already existent environmental problems including soil erosion, deforestation, and land degradation (Ministry of Agriculture, 2011). Natural processes and anthropogenic activities alter the energy budget on the surface and at the main drivers of climate change. Anthropogenic activities have altered the surface of the earth and the composition of the atmosphere. These things have direct and indirect effects on the energy balance of the planet Earth and are responsible for climate change. The year 2020 was known as the year of record temperature and surge in climate-associated disasters such as dots, floods, wildfires, storms, and locust swarms. These phenomena led to the loss of billions of dollars and also caused huge suffering for the ecosystems and communities (Leon et al., 2022).

2 Causes of Climate Change

Any factor that changes the amount of radiation received from the Sun, radiation lost to space, or the energy redistribution within the atmosphere and between the land–atmosphere and ocean is responsible for climate change. The occurrence and intensity of climate change is expected to increase in the near future.

3 Deforestation

Deforestation which involves the cutting down of trees for the use of land for construction, and agricultural purposes makes the soil lose its carbon dioxide absorption capacity and when left to be burnt it emits carbon dioxide. Forests have a great influence on the climatic pattern across the globe as they are involved in the regulation of the water cycle, stability of the atmosphere, and composition of the atmosphere. A significant decline in the area covered over by forest has been observed, approximately 300 million hectares since the year 1990. Approximately 13 million hectares of forests were lost or used for any other purpose between the years 2000 and 2010. This leads to biodiversity degradation and increases 12 to 15% of global warming as this leads to the release of huge amounts of carbon dioxide and impedes the storage of carbon dioxide. Forests are known to cover approximately 42,000,000 square kilometers in temperate, boreal, and tropical lands. They provide ecological, economic, aesthetic, and social benefits to the biotic components, regulate the hydrological cycle, have economic and medicinal value, and act as a hub for a diverse range of plants and animals. Forests have a great impact on climate as they are involved in the exchange of energy, carbon dioxide, etc., in the atmosphere (Feng et al., 2014). Deforestation increases the temperature of the earth's surface, increases the emission of carbon dioxide, depletes the quality of soil, and increases surface run-off which leads to flash floods. Depletion of forest cover changes the regional and global climatic patterns and leads to severe downpours followed by prolonged dry spells. Deforestation is greatly linked with the emission of carbon dioxide because crops that replace trees after land clearance retain less carbon per unit area than forests (Bala et al., 2007). Loss of forest leads to the release of huge quantities of carbon into the atmosphere and it diminishes the carbon removal pathway significantly (Houghton & Nassikas, 2018). Deforestation is the main contributor to climate change. Despite this millions of hectares of forest area are utilized for agricultural purposes every year in order to expand the production of crops (Kunte & Bhat, 2024). Removal of forests is responsible for increasing temperature and reducing evapotranspiration. Forests are known to play a very important role in the maintenance of habitat for biodiversity but the recent research on extinction has confirmed the role of forests in the maintenance of climate that is necessary to support biodiversity. Continuous deforestation could put severe pressure on the remaining forest area by warming and drying regional and local climate. Global forest cover in the year 1900CE was reported to be 5500 million hectares whereas it was only 409 million hectares in the year 2020 (Tarkik-Gautam-Ranjan, 2022).

4 Burning Of Fossil Fuels

Consumption of fossil fuels at the present rate has led to the unprecedented rise of carbon dioxide in the atmosphere over the last 20,000 years. The gaseous emissions released by fossil fuel combustion consist of various toxic air pollutants and carbon dioxide which is much responsible for climate change. Fossil fuel combustion generates the majority of air pollutants across the globe as coal, oil, gasoline, natural gas, and diesel fuel are used for heating, production of electricity, and transportation and have numerous industrial applications. Globally in the year 2011, fossil fuels constituted 82% of the total primary source of energy. Energy-related fossil fuel burning in middle and high-income countries along with biomass burning in low-income nations contributes to the bulk of global air pollution, producing 85% of the airborne inhalable particulate matter and nearly all emissions of nitrogen oxide and sulphur oxides in the atmosphere (US Environmental Protection Agency, 2017). Annually, carbon dioxide emissions surged from 4.85 billion metric tons in the year 1940 to 37.49 billion metric tons in the year 2022 across the globe. The fossil fuel demand is spurring the utilization of heavy, onshore oil and extraction from far-flung areas, tar sands, fracking, and hydraulic fracturing leading to more emissions (Tiseo, 2023). Fossil fuels are the primary catalyst for climate change and a key contributor to greenhouse gas emissions. Climate change and its leading factor, fossil fuel burning are the main causes of avoidable mortality and morbidity across the globe. There is an urgent need to end the reliance on fossil fuels for the well-being of both humanity and the planetary health bracket. The limits should be set on the extraction of fossil fuels to achieve the climate targets. It has been estimated that approximately 50% of fossil methane gas reserves, 33% of oil reserves, and over 80% of the existing coal reserves must remain unextracted by the year 2050 to restrict warming to 2 °C. Since the pre-industrial era, the average temperature of the earth's surface has increased by about 1.1 °C, due to fossil fuel emissions. Fossil fuel combustion has led to air pollution because due to the burning of these fuels, huge large amounts of airborne inhalable particles having a diameter of 2.5 μm or less than 2.5μmand nitrogen oxides, mercury, polycyclic aromatic hydrocarbons, sulphur dioxide, volatile chemicals are released which are responsible for the formation of ground-level ozone (Perera & Nadeau, 2016).

5 Urbanization

During the past few years, the rise in urbanization and alteration of land use patterns has increased the deforestation rate which is responsible for the distortion of climatic patterns across the globe. Urbanization has a profound effect on environmental surroundings and climate throughout the globe and has enhanced the impact of climatic change (Bazrkar et al., 2015). Urbanization is a worldwide phenomenon. Across the globe, only 30% of people were living in urban areas in the year 1950 and this increased to 55% in 2018 and is further expected to be 68% by the year 2050. These substantial transformations have significant effects on populations, ecosystems, and development at regional, community, and worldwide levels (Li et al., 2021). Alterations include contamination of the atmosphere and equity bodies, land utilization, reduction in species diversity, and ecosystem deterioration of which emissions and land management are cheap drivers of climate (Wigginton et al., 2016). The urbanization trends are not uniform and city growth patterns differ distinctly across developing nations. By the year 2050, the number of people living in urban areas will rise by 2.5 billion of which approximately 90% growth will occur in Asia and Africa (DESA, 2018). Various aspects of urbanization exert varying levels of effects on the rise in surface air temperature. For instance, alterations in the energy budget at the surface level, changed wind patterns, electricity generators, industrial operations, and motor vehicles produce substantial amounts of heat in urban areas (Huang et al., 2009). The levels of different airborne pollutants combined with heat retention of non-porous surfaces and scarcity of vegetation cover aggravate the heat in urban areas (Velazquez-Lozada et al., 2006). Urbanization has a negative effect on the local climate. Urbanization is inevitable as it requires accommodating the rising population in cities (Maheshwari et al., 2020). The increase in temperature due to land utilization patterns such as large buildings and expanded impervious surfaces along with continuous consumption of energy in residential, industrial, and commercial areas is known as heat island intensity or urban heat island effect. It is assessed by comparing the temperature anomalies between rural and urban areas (Kataoka et al., 2009). Urbanization and various other associated demands for energy lead to the release of greenhouse gases thereby contributing to changing climatic patterns. The rise in urbanization depicts social development, generally in the developing world but it has posed a substantial risk to the natural environment. As per the Intergovernmental Panel on climate change in the year 2020; urban areas collectively contributed to 75% of the carbon dioxide equivalent emissions. Land in urban areas will triple between the years 2015 and 2050 and the development of new cities will demand huge quantities of raw materials such as wood, metals, sand, etc., leading to a profound impact on ecosystems across the globe. The expansion of urban areas is a great contributor to habitat and biodiversity reduction.

6 Increase in Pollutants

The dramatic escalation in the release of airborne contaminants due to economic and industrial expansion over the last few centuries has made air quality a major environmental concern across the globe. In urban areas, the most predominant components of air pollution are ozone, fine particulate matter, and nitrogen dioxide is more abundant in industrial regions. In urban areas, road traffic is responsible for air pollution. Exhaust released from power plants, cars, and trucks are the major contributors of ambient, nitrogen dioxide which is a catalyst for the formation of photochemical smog in urban areas and industrial areas when combined with sunlight and volatile hydrocarbons. Ozone is released at the ground level involving reactions between hydrocarbons, nitrogen dioxide, and UV radiations. Particulate matter is a combination of inorganic and organic, liquid and solid particles of diverse origins, composition, and sizes. It is a significant contributor to urban atmospheric pollution and has a more significant impact on health (D'Amato et al., 2014). Global warming increases the temperature of the soil and atmosphere and leads to more severe and frequent climatic events such as wildfires, hurricanes, and torrential rains, as a result of this soil erosion occurs which leads to the redistribution of metalloids. The emergence of floods leads to the dispersion of contaminants from industrial areas agricultural and residential lands (Balbus et al., 2013). The second largest carbon reservoir after the oceans is useful in mitigating the effects of climate change. Human activities and environmental systems are dependent upon soil for the availability of water, minerals from organic matter plants, regulation of the hydrological cycle, and carbon storage. Soil pollution alters its carbon sequestration capacity and is responsible for climate change by altering precipitation patterns, temperature, etc. So, the undesirable change in the environment and climate change greatly influence each other through intricate interactions on the surface of the Earth.

7 Volcanic Eruptions

Volcanic eruptions play a significant role in climate change across various time scales. The main component produced from a volcanic eruption is magma which gets released in a solid form on solidifies into huge particles which are called tephra or ash. Volcanic eruption emits gases with CO_2 and N_2 being the most abundant. Explosive volcanic eruption releases sulphur gases into the stratosphere which is the main driver of climatic change (Santer et al., 2015). These eruptions release huge amounts of carbon dioxide into the atmosphere of the earth which is responsible for extreme global warming and brings lethal conditions for the planet Earth. Massive ash fragments descend swiftly from a volcanic eruption. The toxic effects of aerosol clouds trigger reactions within the climatic system. The main effect of volcanic eruption on the climate is to diminish the intensity of solar radiation reaching the surface, decrease the temperature in the troposphere, and alter the circular patterns in the atmosphere. Volcanic activity passes for only a few days. However, the release of volcanic gases and ash affects the climatic pattern for a long period. Once volcanic gases such as water vapor, sulphur, chlorine, carbon, and fluorine get released into the atmosphere they undergo chemical reactions with other compounds and form aerosols and acid particles. Aerosols that remain in the stratosphere led to climate change and have a great influence on the climate for three to four years.

8 Agricultural Expansion

Climate change mitigation and agricultural expansion are closely interconnected and major global challenges. Expansion of agricultural areas and activities results in deforestation and conversion of agricultural lands. Agricultural expansion is a great influence on climate change. There are different mechanisms such as changes in land use and greenhouse gas emissions through which agricultural expansion is responsible for climate change. Agriculture is responsible for the emission of a quarter of approximately 10 to 12 gigatons of carbon dioxide equivalent per year greenhouse gases emission, mainly consisting of nitrous oxide, carbon dioxide, and methane. Stored animal manure and digestion processes in livestock are responsible for the release of methane gas whereas lots of nitrous oxide is released from nitrogenous fertilizers. As per the data given by US EPA, 2019; approximately 9% of the greenhouse gases emitted in the year 2016 were linked to the agricultural sector, particularly rice production, soils, and livestock in the USA. Commercial uses of agricultural land have a negative influence on temperature and local, regional, and global climate. The climate across the globe is getting warmer and greenhouse gases such as methane, nitrous oxide, and carbon dioxide are emitted by numerous human activities including the excessive use of fertilizers and pesticides in agricultural activities (Borrelli et al., 2020). Greenhouse gases released from agricultural fields have huge global warming potential. The intensification of agriculture activities and the potential for further intensification will have a negative influence on non-agriculture aquatic and terrestrial ecosystems across the globe. A two-fold increase in agricultural production has been seen over the past 35 years which is accompanied by a 3.48-fold increase in phosphorus fertilization, a 6.87-fold increase in nitrogen fertilization, a 1.68-fold expansion in irrigated cropland, and a 1.1-fold increase in cultivated land. Reduction in the impact of agricultural expansion on environmental surroundings is one of the main challenges that the whole world is facing during the twenty-first century. Approximately 7 to 10 million square kilometers of forest have been lost over the past few centuries due to the expansion of land under agriculture (Foley et al., 2005).

9 Greenhouse Gases Emission

Continuous rise in the release of greenhouse gases from industrialized countries has led to hydrometeorological events, rising sea levels, and seasonal fluctuations. The weather unpredictability is responsible for food and water scarcity, rising sea levels, and an increase in health crises, thereby affecting the diverse range of transmissions found on the planet Earth. Various sectors are responsible for 60% of the total greenhouse gases emitted which comprises industry, manufacturing, forestry, and energy supply sectors. Global warming arises due to human intervention has a great influence on the ecosystem. As per the report given by the Intergovernmental Panel on Climate Change in the year 2007, the temperature of the earth's surface has escalated notably over the last three decades due to elevated levels of greenhouse gases predominantly from man-made activities. With the rise in anthropogenic activities over the past few years, greenhouse gas emissions across the globe have shown a rise of approximately 70% from 1970 to 2004. The list of greenhouse gases encompasses several components of air contaminants. The rising global average temperature has been observed since the mid-twentieth century most likely as a result of increasing concentration of greenhouse gases released due to various anthropogenic activities. The most important greenhouse gas emitted due to human-induced activities is carbon dioxide and the emission has shown a continuous rise. Increasing greenhouse gas emissions has alterations in the intensity, amount, type, and frequency of rainfall as well as rising extreme events such as droughts,

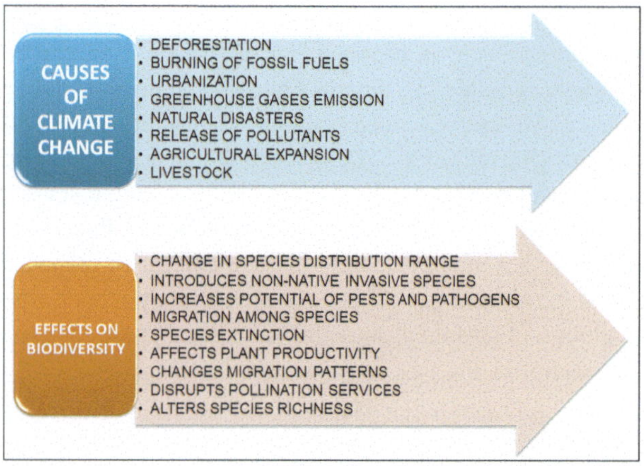

Fig. 1 Causes of climate change and its effects on biodiversity on land

thunderstorms, floods, hurricanes, and heat waves, and these pose a great challenge (D'Amato, 1999). The final byproduct of combustion is the release of carbon dioxide. However, incomplete combustion leads to the production of NO_X and carbon monoxide which further react with other substances present in the atmosphere. The net result of all these photochemical reactions is the formation of ozone, which is another greenhouse gas. Consumption of energy leads to the release of aerosol precursor gases such as sulphur dioxide in the atmosphere; primary aerosols have negative effects on the environment surroundings (Ramanathan & Feng, 2009). Numerous studies have demonstrated that microplastics in soil have a direct contribution to changing climatic conditions by increasing the greenhouse gas emission from the soil into the atmosphere additionally, it promotes soil respiration and has a negative impact on natural carbon sinks such as plants. The causes of climate change and its potential effects on biodiversity on land have been depicted in Fig. 1.

10 Effects on Biodiversity

Biodiversity also referred to as biological diversity includes the various forms of life on the earth's surface varying from microbes to the entire ecosystem. The biodiversity seen at present has been evolving for 4.5 billion years and is greatly influenced by human beings. The wide range of organisms, present in the ecosystem form a web of life on which human beings are much dependent such as water, food, medicine, etc. Most of the human population is dependent upon the forest for their livelihood. However, at present biodiversity is under great threat. Approximately 1,000,000 species are under threat of becoming extinct within a few decades. Forest ecosystems are becoming the sources of carbon instead of carbon sink due to the clearance of forests. 85% of wetland areas such as mangrove swamps and salt marshes, which absorb huge quantities of carbon, have disappeared. Changing climatic conditions modify and enhance anthropogenic disturbances. Climate alterations add up to the prevalence of numerous insect pest species, thereby increasing the loss of crop production. Droughts and floods will have a significant impact on food production. Increased emission of greenhouse gases leads to global warming which will have a negative impact on the pollinators and thus will affect crop productivity as insect pollinators are responsible for the production of crops in 75% of the food crops. Changing climatic conditions decline the diversity and density of pollinators (Shivanna et al., 2020). With an increase or decrease in temperature, the foraging ability of bees decreases thereby adding up to more constraints on the efficiency of pollination in different crop species (Heinrich, 1979). Stubble burning generates huge amounts of toxic gases such as methane, and carbon dioxide, which are responsible for destroying the nutrients and microorganisms present in the soil, and thus affect the fertility of the soil (Abdurrahman et al., 2020).

The threatened species of insects, mollusks, birds, reptiles, amphibians, mammals, fishes, and plants have shown a rising trend in decline over the past few decades (Habibullah et al., 2022). This decline in species is due to various factors such as unsustainable agriculture, loss of habitat, land degradation, pollution, and changing climate. Climate change is the chief driver of biodiversity loss. The increase in temperature and levels of carbon dioxide are deeply associated with climate change. They cause huge changes in the water cycle (evaporation and precipitation) and increase the incidence of extreme weather events such as droughts, floods, and cyclones that lead to the loss of biodiversity. Climate change is the ultimate externality having a negative influence on biodiversity not only through changes in precipitation and temperature but also by influencing the use of land resources, nutrient cycling, etc. Climate change can drive the extinction of more than a million terrestrial species over the next five decades. Diversity loss can be defined as the decline in biodiversity components which comprises biomes, habitats, ecosystems, species, populations, and genetic diverseness. Biodiversity loss is the decline in the variety of life on the planet.

Over the past few years, massive biodiversity loss has been reported which has led to the beginning of the 6th mass extinction crisis because of the environmental changes induced by anthropogenic activities. The effects of climate change on the distribution and abundance of woodland birds in temperate areas. Woodland ecosystems are more prone to climate change due to their complexity and time taken for maturation by trees (Berry et al., 2005). Climate change affects the demography and distribution of disease-causing organisms and the various species of birds because, with alterations in conditions, the rate of reproduction in pathogens and vectors also changes (Faustino et al., 2004). Climate change can increase the risk of extinction of any species because when the precipitation and the temperature exceed the tolerance of any species, it leads to the decline as observed in bumblebee species by Soroye et al., 2020. Rise in temperature influences aspects of growth and development in floral and faunal species. One is the change in species distribution range and the other is the change in phenological events. Animal and plant species have become adapted to their original habitat conditions over thousands of years. With the increase in temperature due to changing climatic conditions, the species shift towards higher altitudes and polewards to find suitable temperatures for growth and development. The timing of plant activity has shifted with the global environmental changes. Various species are not able to cope with changing environmental conditions and thereby lag which leads to their mass extinction (Shivanna, 2022). Long-term observations over 100 years have depicted that various species of bumblebees in Europe and North America are not able to cope with changing climatic conditions and therefore perish away from southern parts of their range (Kerr et al., 2015). Various flowering plants are dependent upon animals for dispersal of seeds. Defaunation caused by due to changing climatic conditions and various other environmental disturbances has declined the long-distance dispersal of seeds (Fricke et al., 2022). Changes in species distribution due to climatic alteration have declined the synchronization between the availability of pollinator species and the time of flowering in plant species. When any plant species migrates the pollinator species will not be able to migrate and similarly when many pollinator species migrate, plant species on which sustenance is dependent will not migrate. Disruption in pollination services due to alterations in climatic patterns has been reported (Gerard et al., 2020).

Agricultural land has led to the loss of tropical forest which in turn has severe negative impacts on the tropical biodiversity because habitat destruction of any species threatens its survival. The increase in deforestation rate has made the remote areas of forest more accessible thereby increasing the pressure of hunting and trapping and various other kinds of anthropogenic activities. Deforestation at a very rapid level has made the survival of native animal species difficult and they are perishing continuously, therefore becoming the cause of extinction of various species (Sahney et al., 2010). Deforestation also destroys the microbial communities that aid in nutrient cycling, clean water production, and removal of pollutants. It has been estimated that due to rainforest deforestation approximately 137 species of plants, animals, and insects have been lost every single day. The rate of rise in global mean temperature is much higher than any other natural warming over the past 18,000 years, which is indicative of the fact that various species will struggle to shift their habitats rapidly enough to adapt to a quickly changing climate. Global climate change is expected to bring certain changes in vegetation cover. With the increase in temperature of the earth's surface species are expected to migrate to higher latitudes and altitudes. Certain regions of endemic and tropical alpine vegetation might be replaced by montane cloud forests which are currently found at lower altitudes (Halpin, 1994).

Natural disasters such as the eruption of Krakatau volcano in Indonesia have affected the floral and floral species of the islands located near Krakatau, the southeastern region of Sumatra, and the western part of Java (Darnaedi & Zulkarnaen, 2017). Natural disasters such as landslides, volcanic eruptions, wildfires, droughts, and earthquakes have devastating impacts on biodiversity. Studies conducted by Villen-Perez et al. (2020) have revealed that in Finland climate change has influenced the plant species. The temperature-sensitive dwarf shrubs, grasses, herbs, lichens, and bryophytes shift northwards to cooler climatic regions. Climate change will disrupt the natural communities and cause species extinction in the near future. Changing climate variables have caused modifications in the physiology, behavior, and morphology of species. For instance, the body weight of *Neotema* sps. has decreased with the rise in temperature over the past eight years. In Scotland, *Cervus elaphus* shows faster growth in warmer springs thereby increasing the adult body size. Eliño during the years 1997–1998 had a huge impact on land ecosystems across the globe. Anthropogenic activities such as alterations in land use patterns, ecosystem degradation, species exploitation, and introduction of invasive species have increased the impacts of climate change.

Climate change leads to migration and extinction of animals. Because of the changing climatic conditions, the Fennec fox, a migratory mammalian species searches for harsh environmental conditions for its survival. A *Hubara bustard*, a bird species is under serious threat of extinction. Climate change alters the migratory patterns in migratory birds that utilize seasonal wetlands. This increases the conflicts between large mammals and humans, especially in areas where there is low rainfall. Changes alter genetic structure and rate of breeding in different populations also in the duration and intensity of dry and rainy seasons (Thirgood et al., 2004). Migration of species, alteration in the composition of species, chains, and availability of resources, disease spread to new areas, and alterations in features of protected areas are the major impacts of climate change on biodiversity. Climatic alteration affects the behavior and growth of organisms, modifies the age structure and size of the population, and affects the nutrient cycling, flow of water, composition of species, and interactions among them. Indirectly, climate change induces alterations in the disturbance regime of a region. Climate change changes the timing of various biological events such as growth, emergence, and reproduction in a few invertebrate species due to warmer conditions in the autumn and spring seasons. The growth of shrubs and grasses is dependent upon the duration and intensity of rainfall. Some plant species have shallow and dense root systems and they absorb moisture from water present in the upper layer of the soil and less precipitation due to climate change reduces the resilience of such plant species (Chidumayo, 2005). A rise in temperature by 38 °C increases the evapotranspiration due to heat conditions that lead to the drying up of the soil. Therefore, warmer condition decreases the productivity of plants (Vanacker et al., 2005). The rapid increase in temperature due to climate change disrupts the interconnectedness among species, leading to the transformation of existing communities and a shift in the movement of species. This will be responsible for the species extinction (Malcolm et al., 2006). Climate change has a significant influence on species richness. Climate change has caused the extinction of various species in recent times. Many species have shifted their ranges and have migrated polewards and upwards in elevation during the last century and this continues to occur. Local communities are dispersing and incorporating a greater number of warm-adapted species.

Reptiles are dependent upon external environment temperature for the optimization of their body temperature which influences their physiology, behavior, and reproduction. Continued climate change poses a serious risk for populations of reptiles. Changes in atmospheric temperature alter the temperature of their body and their physiological performance which makes them vulnerable to extinction. An increase in air temperature increases the temperature inside the nests of lizards and recent studies have proven that over the next century, a significant decline will be seen in the lizard populations. Changing climatic conditions has led to the extinction of different animal species. Climate change is responsible for the extinction of amphibian species such as the Monteverde harlequin frog and golden toad (McCarthy et al., 2001).

Variations in the temperature, precipitation, and increase in frequency and intensity of meteorological hazards such as fires, cyclones, storms, droughts, floods, heat waves, and epidemics have a great influence on climate change. As per the Intergovernmental Panel on Climate Change (IPCC), by the year 2100, an increase of 1.4 to 5.8 °C in the average temperature has been predicted across the globe. Forests protect biodiversity, regulate the water cycle and atmospheric temperature, preserve soil cover, and aid in mitigating the impact of climate change. With the increase in urbanization, the deforestation rate has also increased and this has led to the destruction of natural climatic patterns and amplifying the catastrophic consequences of natural hazards. The inability to stabilize the climate poses a huge threat to biodiversity. There is an urgent need to develop an understanding of the effect of urbanization on the local environment in terms of rainfall, temperature, and evaporation. Urbanization and climate change are the prominent drivers of change in biodiversity and ecosystems. There is a direct link between pollution and climate crisis as they constitute the complex struggles of the twenty-first century. Approximately 2000 years ago, it was pointed out that the volcanic eruption of Mount Etna in 44 BC decreased the sunlight which led to the cooling and was responsible for the weathering of crops, leading to famine in Egypt and Rome.

Recent advancements in agriculture activities have made a significant contribution to climate change and this has led to the loss of biodiversity and land degradation. Conversion of land into agricultural fields leads to the destruction of the habitat of various plant and animal species, which become more prone to extinction. Climate change has altered the terrestrial, marine, and freshwater ecosystems. It has declined the population of local species, enhanced the incidence of various diseases, and led to the mass mortality of different floral and floral species thereby responsible for climate-induced extinction.

11 Conclusion

There is a critical need for policy measures that go beyond the preservation of forest cover to safeguard the biodiversity of tropical forest ecosystems. Achievement of sustainable development and maintenance of situational awareness are essential for reducing damage caused by climatic change. The evaluation of the influence of climatic change on the flora and fauna species is challenging because changes happen gradually and the impact of climate change interacts with the already prevailing environmental stresses. Species which are much more vulnerable to changing climatic conditions are used as indicator species as they did in evaluating the sensitivity of the climate of the whole ecosystem. But effective conservation of biodiversity, a nation should investigate wealth and efforts in reducing emission of air pollution, and greenhouse gases to prevent global warming and other natural disasters. Disruption of ecosystem processes such as carbon dioxide fixation, respiration, decomposition, and strength due to loss of species can affect the functioning of the whole ecosystem. So, the issue of climate-induced loss of biodiversity is of significant concern, and enhancing our understanding of the negative influences of climate change diversity, and developing the methods to reduce the effects and decrease the anthropogenic activities are crucial for limiting the damage. Without protecting biological diversity and reducing human-induced activities, the achievement of sustainable development becomes unattainable.

References

Abdurrahman, M. I., Chaki, S., & Saini, G. (2020). Stubble burning: Effects on health & environment, regulations and management practices. *Environmental Advances, 2*, 10001.

AR6 Climate Change. (2021). *The sixth assessment report on climate change*. IPCC

Bala, G., Caldeira, K., Wickett, M., Phillips, T. J., Lobell, D. B., Delire, C., et al. (2007). Combined climate and carbon-cycle effects of large-scale deforestation. *Proceedings of the National Academy of Sciences of the United States of America, 104*(16), 6550–6555.

Balbus, J. M., Boxall, A. B. A., Fenske, R. A., McKone, T. E., & Zeise, L. (2013). Implications of global climate change for the assessment and management of human health risks of chemicals in the natural environment. *Environmental Toxicology and Chemistry, 32*, 62–78.

Bazrkar, M. H., Zamani, N., Eslamian, S., Eslamian, A., & Dehghan, Z. (2015). Urbanization and climate change. In W. Leal Filho (Ed.), *Handbook of climate change adaptation*. Springer. https://doi.org/10.1007/978-3-642-38670-1_90

Berry, P. M., Harrison, P. A., Dawson, T. P., & Walmsley, C. A. (Eds.). (2005). *Monarch 2: Modelling natural resource responses to climate change*. UKCIP Technical report.

Borrelli, P., Robinson, D. A., Panagos, P., Lugato, E., et al. (2020). Land use and climate change impacts on global soil erosion by water (2015–2070). *Proceedings of the National Academy of Sciences, 117*(36), 21994–22001.

Chidumayo, E. N. (2005). Effects of climate on the growth of exotic and indigenous trees in central Zambia. *Journal of Biogeography, 32*, 111–120.

D'Amato, G. (1999). Outdoor air pollution in urban areas and allergic respiratory diseases. *Monaldi Archives for Chest Disease, 54*, 470–474.

D'Amato, G., Bergmann, K. C., Cecchi, L., Annesi-Maesano, I., Sanduzzi, A., Liccardi, G., Vitale, C., Stanziola, A., & D'Amato, M. (2014). Climate change and air pollution: Effects on pollen allergy and other allergic respiratory diseases. *Allergo Journal International, 23*(1), 17–23.

Darnaedi, D., & Zulkarnaen, I. (2017). Disasters and biodiversity in the ring of fire of the Indonesian archipelago. *Biology International Special Issue, 36*, 39–46.

DESA. (2018). World Urbanization Prospects: The 2018 Revision (ST/ESA/SER.A/420). United Nations: New York, NY, USA.

EPA. (2019). Global greenhouse gas emissions data. Retrieved 14 Feb from https://www.epa.gov/ghgemissions/sources-greenhouse-gas-emissions#agriculture

Faustino, C., Jennelle, C. S., Connolly, V., Davis, A. K., Swarthout, E. C., Cooch, E. G., & Dhondt, A. A. (2004). *Mycoplasma gallisepticum* infection dynamics in a house finch population: Empirical analysis of seasonal variation in survival, encounter and transmission rate. *Journal of Animal Ecology, 73*, 651–669.

Feng, S., Hu, Q., Huang, W., Ho, C. H., Li, R., & Tang, Z. (2014). Projected climate regime shift under future global warming from multi-model, multi-scenario CMIP5 simulations. *Global and Planetary Change, 112*, 4152. https://doi.org/10.1016/j.gloplacha.2013.11.002.

Foley, J. A., DeFries, R., Asner, G. P., Barford, C., Bonan, G., Carpenter, S. R., Chapin, F. S., Coe, M. T., Daily, G. C., Gibbs, H. K., Helkowski, J. H., Holloway, T., Howard, E. A., Kucharik, C. J., Monfreda, C., Patz, J. A., Prentice, I. C., Ramankutty, N., & Snyder, P. K. (2005). Global consequences of land use. *Science, 309*(5734), 570–574.

Fricke, E. C., Ordonez, A., Rogers, H. S., et al. (2022). The effects of defaunation on plants' capacity to track climate change. *Science, 375*, 210–214.

Gerard, M., Vanderplanck, M., Wood, T., & Michez, D. (2020). Global warming and plant-pollinator mismatches. *Emerging Topics in Life Sciences, 4*, 77–86.

Habibullah, M. S., Din, B. H., Tan, S. H., & Zahid, H. (2022). Impact of climate change on biodiversity loss: Global evidence. *Environmental Science and Pollution Research, 29*, 1073–1086.

Halpin, P. N. (1994). Latitudinal variation in the potential response of mountain ecosystems to climate change. In M. Beniston (Ed.), *Mountain environments in changing climates* (pp. 180–203).

Heinrich, B. (1979). Keeping a cool head: Honeybee thermoregulation. *Science, 205*, 1269–1271.

Houghton, R. A., & Nassikas, A. A. (2018). Negative emissions from stopping deforestation and forest degradation, globally. *Global Change Biology, 24*, 350–359

Huang, S., Taniguchi, M., Yamano, M., & Wang, C. H. (2009). Detecting urbanization effects on surface and subsurface thermal environment–a case study of Osaka. *Science of the Total Environment, 407*(9), 3142–3152.

Huang, Y., Haseeb, M., Usman, M., & Ozturk, I. (2022). Dynamic association between ICT, renewable energy, economic complexity and ecological footprint: Is there any difference between E-7 (developing) and G-7 (developed) countries? *Technology in Society, 68*, Article 101853.

Hussain, M., Liu, G., Yousaf, B., Ahmed, R., Uzma, F., Ali, M. U., & Butt, A. R. (2018). Regional and sectoral assessment on climate-change in Pakistan: Social norms and indigenous perceptions on climate change adaptation and mitigation in relation to global context. *Journal of Cleaner Production, 200*, 791–808.

Kataoka, K., Matsumoto, F., Ichinose, T., & Taniguchi, M. (2009). Urban warming trends in several large Asian cities over the last 100 years. *Science of the Total Environment, 407*(9), 3112–3119.

Kerr, J. T., Pinder, A., Galpern, P., et al. (2015). Climate impacts on bumblebees coverage across continents. *Science, 349*, 177–180.

Kunte, G., & Bhat, V. (2024). Deforestation, climate change and the sustainability of agriculture: A review. *Journal of Resources and Ecology, 15*(1), 140–150.

Leon, M., Cornejo, G., Calderón, M., González-Carrión, E,. & Florez, H. (2022). Effect of deforestation on climate change: A co-integration and causality approach with time series. *Sustainability, 14*.

Li, G., Fang, C., & Qi, W. (2021). Different effects of human settlements changes on landscape fragmentation in China: Evidence from grid cell. *Ecological Indicators, 129*, 107927

Maheshwari, B., Pinto, U., Akbar, S., & Fahey, P. (2020). Is urbanisation also the culprit of climate change?–Evidence from Australian cities. *Urban Climate, 31*, Article 100581.

Malcolm, J. R., Liu, C., Neilson, R. P., Hansen, L., & Hannah, L. (2006). Global warming and extinctions of endemic species from biodiversity hotspots. *Conservation Biology, 20*, 538–548.

McCarthy, J. J., Canziani, O. F., Leary, N. A., & DokkenDJ, W. K. S. (2001). *Climate change 2001: Impacts, adaptation, and vulnerability*. Cambridge University Press, UK.

Mihiretu, A., Okoyo, E. N., & Lemma, T. (2021). Awareness of climate change and its associated risks jointly explain context-specific adaptation in the Arid-tropics. *Northeast Ethiopia SN Social Sciences, 1*(2), 1–18.

Ministry of Agriculture. (2011). *Agriculture sector programme of plan on adaptation to climate change*. Addis Ababa, Ethiopia

Perera, F., & Nadeau, K. (2016). Climate change, fossil-fuel pollution, and children's health. Fossil-fuel pollution and climate change.

Ramanathan, V., & Feng, Y. (2009). Air pollution, greenhouse gases and climate change: Global and regional perspectives. *Atmospheric Environment, 43*, 37–50.

Sahney, S., Benton, M. J., & Falcon-Lang, H. J. (2010). Rainforest collapse triggered Pennsylvanian tetrapod diversification in Euramerica. Geology. https://www.academia.edu/368820/Rainforest_collapse_triggered_Pennsylvanian_tetrapod_diversification_in_Euramerica

Santer, B. D., Solomon, S., Bonfils, C., Zelinka, M. D., Painter, J. F., Beltran, F., Fyfe, J. C., Johannesson, G., Mears, C., Ridley, D. A., Vernier, J. P., & Wentz, F. J. (2015). Observed multivariable signals of late 20thand early 21st century volcanic activity Geophys. *Research Letters, 42*, 500–509.

Sarkar, P., Debnath, N., & Reang, D. (2021). Coupled human-environment system amid COVID-19 crisis: A conceptual model to understand the nexus. *Science of the Total Environment, 753*, Article 141757.

Schuurmans, C. (2021). *The world heat budget: expected changes* (pp. 1–15). Climate Change CRC Press.

Shivanna, K. R. (2022). Climate change and its impact on biodiversity and human welfare. *Proceedings of the Indian National Science Academy, 88*(2), 160–171.

Shivanna, K. R., Tandon, R., & Koul, M. (2020). Global pollinator crisis and its impact on crop productivity and sustenance of biodiversity. In R. Tandon, K. R. Shivanna, & M. Koul (Eds.), *Reproductive ecology of flowering plants: Patterns and processes* (pp. 395–413). Springer.

Soroye, P., Newbold, T., & Kerr, J. (2020). Climate change contributes to widespread declines among bumble bees across continents. *Science, 367*(6478), 685–688.

Tarkik-Gautam-Ranjan. (2022). Global deforestation and its relation to animal extinction. *World Journal of Advanced Research and Reviews, 15*(01), 499–511.

Thirgood, S., Mosser, A., Tham, S., Hopcraft, G., Mwangomo, E., Mlengeya, T., et al. (2004). Can parks protect migratory ungulates? The case of the Serengeti wildebeest. *Animal Conservation, 7*, 113–120.

Thurman, H. V., & Elizabeth, A. B. (2001). *Introductory oceanography*. Prentice Hall, pp 1–352.

Tiseo, I. (2023). *Annual carbon dioxide (CO_2) emissions worldwide from 1940 to 2022*. Statista: Energy and Environment-Emissions

US Environmental Protection Agency. (2017). *Greenhouse gas emissions*. https://www3.epa.gov/climatechange/ghgemissions/

Vanacker, V., Linderman, M., Lupo, F., Flasse, S., & Lambin, E. (2005). Impact of short-term rainfall fluctuation on interannual land cover change in sub Saharan Africa. *Global Ecology and Biogeography, 14*, 123–135.

Velazquez-Lozada, A., Gonzalez, J. E., & Winter, A. (2006). Urban heat island effect analysis for San Juan, Puerto Rico. . *Atmospheric Environment, 40*(9), 1731–1741.

Villen-Perez, S., Heikkinen, J., Salemaa, M., & Makipaa, R. (2020). Global warming will affect the maximum potential abundance of boreal plant species. *Ecography, 43*, 1–11.

Wigginton, N. S., Fahrenkamp-Uppenbrink, J., Wible, B., & Malakoff, D. (2016). Cities are the Future. *Science, 352*(6288), 904–905.

WMO. (2006). *WMO global ozone research and monitoring project reports*.

Influence of Cropping System on Soil Properties

Suganthi Varathiyan, Dhanalakshmi Thiyagarajan, Subakanmani Swaminathan, Kerenhap Jyotsna Sweety, and Saranya Harikumar

Abstract

Soil is a very precious resource and a very complicated ecosystem. Deplorably a variety of invasive anthropogenic activities in intensive agriculture have rapidly degraded soil globally and continue to do so, with antagonist effect for ecosystem and human health. This raises concerns because, soil is a non-renewable resource with in a human time frame. Due to vanishing of agricultural land brought about by urbanization, there is currently a need to grow more crops in limited space. Thus, soil fertility is progressively declined. Maintaining soil fertility has become a significant challenge in today's world. Different management techniques are employed nowadays to preserve soil fertility. This review aids in comprehending how cropping systems preserve soil fertility and enhance crop yield. Soil quality needs to be preserved and enriched in continuous cropping systems in order to support agricultural productivity and environmental conservation for future generation.

Keywords

Soil health · Cropping system · Sustainable agriculture · Conservation · Fertility

1 Introduction

Soil is the building block of agriculture, providing critical physical, chemical and biological support for plant growth and ecosystem function. Maintaining Soil fertility is a significant concern in today's world (Heather et al., 2024). Crops require fertile terrain for optimal yields. Agricultural land is vulnerable to various challenges, including physical, chemical and biological degradation. These factors include erosion, compaction, salinization soil contamination, organic matter loss and biodiversity loss (Crotty et al., 2022). Industrialized farming practices utilize specialized machinery to apply inputs, but this leads to unsustainable cropping system, resulting in soil fertility loss, soil compaction and hinder root growth. While increased cultivation and climate can deplete soil organic matter, narrow rotation can reduce biodiversity and increase the risk of soil-borne disease. Protecting and improving soil health is critical for long-term agricultural success and food security (Tahat, 2020). Inadequate clearance procedures, improper land use, overgrazing and overexploitation have led to wind and water erosion, salinization, compaction and severe nutrient depletion. Kopittke et al. (2019) the affected region is still mounting and most of the land surface is depreciated.

Cropping system, a crucial aspect of farming, is the pattern of crops utilized on a farm and how they interact with resources, other farms and technology. To improve soil productivity and fertility, agricultural systems should address deficiencies in macro, micro and secondary nutrients that have created a constraint in increasing the sustainability and productivity of soil (Khanday et al., 2017). To address nutritional deficits, research should focus on crop distribution, availability, solubility and uptake. Chemical fractions, soil pH, soil organic matter, microbial activity, management techniques and soil–plant interaction all influence soil nutrient content (Agarwal et al., 2016). According to

S. Varathiyan (✉) · D. Thiyagarajan · K. J. Sweety · S. Harikumar
Nehru Institute of Health Sciences, Coimbatore, Tamil Nadu, India
e-mail: sk97970@gmail.com

S. Swaminathan
Nehru Arts and Science College, Coimbatore, Tamil Nadu, India

J. A. Parray (ed.), *Soil and Land Use Change*, Sustainable Landscape Planning and Natural Resources Management,
https://doi.org/10.1007/978-3-031-93075-1_8

the Intergovernmental Panel on Climate Change (Ginevra), greenhouse gas emissions from modern agricultural practices contribute to climate change, accounting for approximately 24% of total anthropogenic emission. The chemical and physical attributes of soil are crucial for the movement and reaction of water, solutes, and gases in soils. Understanding these properties is crucial for planning appropriate agricultural practices and for knowing the behavior of soil under stress. The Earth's surface has changes due to human-induced changes in land use, which have impacted soil physical chemical characteristics such as soil fertility, soil erosion sensitivity and moisture content of soil. These changes could be attributed to soil compaction, and it diminishes the yield of soil and ecological integrity by decreasing soil volume. Soil physical and chemical parameters have been recommended as appropriate markers for determining the impact of land use changes and management.

Cropping systems have a significant impact on soil biota by providing plant residual food sources and affecting the physical and chemical environment. Crop type, intensity, tillage, residue management and fertilizer all influence soil biological activity. This paper reviews how cropping system affects soil health, soil organic matter, soil nutrient level, soil biodiversity and microbiological indicators caused by anthropogenic activities in agriculture. The findings can be used to optimize cropping system in the future.

2 Cropping System and Soil Properties

The precise placement and management of crops all over time in an agriculture field is referred to as a cropping system. The primary objective of cropping systems was to optimize agro-system productivity; however, in contemporary agriculture, cropping system environ mental sustainability has received more attention. It encompasses a number of variable, such as the kinds of crops cultivated, their placement, spacing, and management techniques used. There are many different cropping systems based on climate, soil type and resource availability. According to the five essential function evaluation standards—nutrient cycling, water relations, biodiversity and habitat and buffering, and physical stability and support the aim of maintaining soil health is to guarantee the long-term stable high productivity and environmental sustainability of cropping systems (Harfield et al., 2017). Cropping systems have a variety of geographical and temporal effects on soil health and quality. These include crop rotation, intercropping and related agronomic techniques.

2.1 Impact of Crop Rotation on Soil Physical and Chemical Health

Crop rotation has long been recognized as a useful strategy to sustain the biodiversity of agro ecosystems by boosting soil health, suppressing diseases and insect's outbreaks, and ultimately raising yields (Barbieri et al., 2019). Crop varieties employed in rotation, rotation series and the frequency of application of certain crops (Bainard et al., 2017a, 2017b) are some of the elements that affect the value and efficiency of a crop rotation. When an ecological farming system was compared to a standard farming system, the ecological system exhibited better physical in addition chemical characteristics of the Earth as well as higher quality organic matter. The healthy soil is primarily the result of circular processes, for example, a healthy soil promotes crop growth, which increases the return of plant residues to the soil and the activity of the root systems, both of which improve soil structure. The physical qualities of healthy soil, which include high rooting depth, ability to hold water, as well as structure, are also influenced by appropriate soil biology. The foundation of sound physical qualities in soil is organic matter in the soil, which also serves as the basis for nearly all other measures related to soil health. There was a relationship involving tillage along with rotation, following no-till farming, soil organic carbon had 7% more than it was during kernel tillage. Two long-term rotation trails—continuous corn, continual soybean, corn-soybean, as well as corn-soybean-wheat—were illustrated this. The most diverse rotation of corn-soybean–wheat enhanced soil organic carbon content compared to continuous soybean. When an ecological farming system was compared to a standard farming system, the ecological framework exhibited better soil physical and chemical properties as well as better quality organic material. The standard farming system's chemical inputs were substituted by multifunctional crop rotation, ecological plant nutrition and mechanical and physical weed control. Cultivated tap-rooted crops into a crop rotation system may facilitate the establishment of more roots in deeply buried soil layers, hence providing a higher degree of accessibility to subsurface nutrients and water. Increased Carbon: Nitrogen ratio crops in their residues contribute to increased carbon and nitrogen levels in the upper 20 cm of the soil, according to Zuber et al. (2018). Alternatively, excess nitrogen is unable to be eliminated in large amounts on its own, rather it needs to be present in the soil in enough quantities to create organic matter which can store the nitrogen The degree of nitrogen fixation is determined by the following factors, soil temperature; soil moisture level; the presence or inoculation of the appropriate symbiotic bacteria; the amount of photosynthesis, which provides the symbiotic bacteria with carbon and energy; and the length of the legume phase. After combining an integrated and sustainable rotation, greater quantities of biological nitrogen and nitrification were discovered in the ecological framework. The versatile crop rotation, environmental plant nutrient management, and kinetic and physical weed management in the natural landscape took the place of the chemical interventions in the integrated system.

Intercropping system effects on the soil

With sustainable intensification in the context of the changing climate, legume intercropping has huge potential to feed the world's growing population (Jhariya et al. 2021). Worldwide, intercropping has been a long-standing practice. When compared to monocropping, intercropping enhanced the organic carbon content of the soil by 4%, who used a C/N analyzer treating the soil with 0.5 M HCL. In a sandy soil with little organic matter, Intercropping systems, which include maize, wheat and faba beans, had estimated 23, 4 and 11% higher root biomass as well as organic C and N content in the top 20 cm soil layer when compared to the species in rotation. Cultivating two or more crops concurrently on the same plot of land at the same is a significant agronomic practice (Kumar et al., 2020). Legume intercropping raises system profitability and productivity per unit of land. 15–20% of the world's food needs are met through intercropping. By using less chemical inputs to produce food that is high in protein, minerals and vitamins, it establishes an environmentally sound agricultural production system.

Impact of cover cropping system on soil health

A crop which is planted primarily to improve and preserve the soil as opposed to yield food is considered a cover crop, as it contributes to the ecological sustainability of the cropping system. In order to minimize weed infestation and maintain a consistent production, cover crops could be required. Numerous studies have shown that growing cover groups can dramatically raise soil organic carbon and total nitrogen, particularly in diminished tillage regimes. Weed reduction, biological Nitrogen fixation, preventing soil erosion and soil penetration are some of the specific objectives of cover crops. Leguminous Cover crops fix 60–80 kg ha^{-1} of nitrogen into the soil as compared to non-leguminous cover crops. Additionally, non-leguminous cover crops can minimize nitrogen loss by immobilizing excess soil nitrogen.

Cover crops can improve soil porosity through the development of bio-pores in the soil through the growth of roots and encouraging earthworm activity, in addition to reducing runoff as well as boosting water infiltration into soil. When compared to no cover crops, cover crops have been demonstrated to improve soil physical attributes by reducing soil bulk density. Enhancing organic carbon content in soil and aggregation boosting the percentage of macropores (Cercioglu et al., 2018), enhancing water retention, saturated hydraulic conductivity, improving water infiltration alongside dropping soil loss (Haruna et al., 2018).

In order to increase soil fertility, cover crops were planted as green manures, acting as a mulch and soil supplement (Kasper & Singer, 2011). Because of the increased activity of microbes, incorporating cover crops to the crop rotation cycle could lead to early boosts in mineralization and CO_2 emissions. Microbial immobilization rises with time, leading to an upsurge in soil organic carbon (Haruna et al., 2018). Cultivation of cover crops, including leguminous along with non-leguminous, significantly increases soil carbon content at a loamy deepness of 0–30 cm. Enhanced soil eater may be result of the biospore created by cover crops roots. Nouri et al. (2019) demonstrated that on a silt loam soil, the application of vetch and wheat cover crops enhanced cumulative infiltration by 86 and 116%, accordingly as compared to non-cover corps under a no till management practice.

3 Soil Properties as Influenced by Soil Organic Matter

In order to maintain an appropriate soil structure, control the thermal regime or manage the water dynamics in the soil, among other process that improve soil physical qualities, the soil organic matter is essential. The SOM also has a substantial impact on biochemical processes such as nutrient cycling, the binding and breakdown of dangerous chemicals and elements (Yang, 2019) as well as soil biological characteristics (Gong et al. 2021). The high SOM concentration boosts both nutrient supply and soil buffering ability (Shi et al., 2019). Increasing SOM content also contribute to increase C sequestration in soil (LoLorenz & Lal, 2018). As a result, erosion losses of topsoil are reduced. As a result, preserving or increasing SOM content is critical for agro ecosystem productivity and assuring food safety in the future (Stavi et al., 2016). Organic fertilizers, compost and digestate can enhance soil organic matter and provide nutrients (Baştabak & Koçar, 2020). For the agricultural ecosystem to be sustainable over the run, soil organic matter retention is essential. By distributing and causing leaf litter, a cropping system that includes legume other deep-rooted crops can upend soil organic matter, improving the chemical, biological and physical properties of the soil (Pries et al., 2018). Long-term cultivation progresses the soils SOC and nutrient condition In contrast, a significant portion of the crop gets discarded as straw and cereal grains in the rice- wheat combination, leading to a significant loss of nutrients, the agro- forestry method also enriches the soil with organic matter. Due to addition of micronutrients from leaf litter and roof turnover, poplar-based agroforestry raised overall Zn, Cu, Fe and Mn levels relative to the rice-what system (Dhaliwal et al., 2021).

4 Influence of Soil Compaction on Soil Property

The process of rearranging soil grains to reduce vacant space and bring them into close contact with one another, hence increasing the bulk density is known as soil compaction. The biological, chemical as well as physical process in soil process are influenced by soil compaction, which is contingent upon soil structure. Topmost soil compaction predominantly impacts the foremost layer of the soil in farmed areas. While compaction of subsoil is also found at a certain depth. With a few exceptions, where a small amount of soil compaction can offer advantages for specific soil types, particularly sandy soils. The soil texture affects the soil, susceptibility to compaction. Sandy soils are only marginally vulnerable to soil compaction, conversely, moderate or extremely fine loamy and clayey soils that have modest water content are less susceptible than silt loams with lower colloid levels.

Rebound effects suggest that increases soil organic matter may decrease compatibility by enhancing elasticity and/or resistance to deformation. Clayey and silty soils with high organic carbon contents may even become less compactible at high moisture levels. The percentage of water in the soil has a major impact on the compaction process of the soil. It influences the maximum allowable ground pressure in the soil, as well as the penetration resistance and support capability. The clay content and mineralogical feature of a soil also affect its susceptibility to compaction at a given soil moisture and energy level. In general, a soil that has very little moisture content is less prone to compaction than one that has a lot of moisture. Nevertheless, the soil loses its compressibility when the moisture level is so high that water occupies every pore in the soil. A soil's dry bulk density and cone index rise with an increase in machine passing frequency, leading to top soil compaction and unfavorable physical soil conditions for seed emergence or early passages accounts for a significant amount of the overall compaction of the soil and 10 passes can impact the soil as deep 50 cm. Natural causes of soil compaction are confined to the ground's topmost 5 cm. Urban pressure and trampling on a site can compact the soil up to 20 cm, while mechanical activities can compact the soil up to 60 cm. the soil composition, regardless of origin, affects a variety of processes, including plant growth Soil erosion pesticide diffusion and water dynamics.

Compacted soil has accessible nitrogen and crops use nitrogen less efficiently, which can lead to elevate fertilizer demands. According to reports, compaction of the soil, which eventually results in higher water contents and denitrification process, appears to reduce NOx emissions from the soil, but it additional enhances ammonia volatilization when compared to un compacted soils. The anaerobic soil conditions that are favored by soil compaction might lead to a decrease in methanotropic bacteria and increase in methanogenic bacteria. Compacting well-drained soils could decrease the rate of destruction of CH4 by up to 58%.

4.1 Compaction on the Soil and Its Impact on Soil Biodiversity

The way that soil compaction impacts the chemical and physical attributes of the soil, and eventually the soil ecosystems, is determined by changes in the physical properties of the soil. The sort of soil, temperature, and the degree of compaction, and inversely, may impact the advantages of compaction for the diversity of organisms in the soil. The total biomass of soil microorganisms is adversely affected by soil compaction. Soil compacting resulted in a 13%–36% reduction in air-filled porosity, it which diminished soil aeration and, ultimately, the biomass of microbial as well as carbon. Soil physical qualities are vital in assessing its suitability for agricultural, environmental and engineering purposes. Excessive compactness can negatively impact the root environment, reduce water penetration and increase runoff and erosion (Mudarisov et al., 2020).

Biological, chemical and physical soil factors can all be combined to measure the quality of the soil. Depending on the soil system, microbial activity reveals the microbiological process carried out by soil microorganism. The most prevalent types of microorganism in soil are fungi and bacteria, which are crucial to the rates at which trash decomposes and nutrients are transformed. The organization of microbial communities is largely determined by several factors, including pH, cation exchange capacity and the concentration of organic matter and soil texture (Puissant et al., 2021). By giving some microorganisms a good place to live, these features have the direct ability to change the structure of microbial communities and causes a maximal destruction process. The microbial population in soil and their efficiency in the process of breaking down plant waste may be influence by the relationships among soil organic matter, total nitrogen levels and soil texture. The decline in phosphorus-containing microbial biomass subsequent soil compaction was also documented. Enzymatic activity in the soil can be impacted by any stress or disturbance. Compaction of the soil influences its physical and chemical characteristics, which leads to decrease in the activities of phosphatase, urease, and dehydrogenase. However, there are also reports of an increase in phosphatase activity. Compacted soils have lower eukaryotic/ prokaryotic ratios, more iron and sulfate reducers, and higher methanogens than un-compacted soils due to anoxic conditions in the soil customize the microbial community and endorse organisms able to survive under these condition. Several variables may affect the soil-microbe-plant complex in cropping system, altering the manner in which it functions. The composition as well as adaptability of soil microbial communities can be altered by a variety of factors, including soil type, moisture and temperature, organic carbon content, pH, electrical conductivity and calcium level (Bainard et al., 2017a, 2017b). The symbiosis of various soil microbes in cropping systems provides several advantages to crop vegetation (Fierer, 2017). Specifically, symbioses with functional microorganisms such as mycorrhizal fungi, beneficial endophytic fungi, are plant growth promoting bacteria are essential to many biofunctions of agro-ecosystem.

5 Conclusion

Refinement of soil health content and advancement of new assessment standards for soil health and quality are noteworthy accomplishment. These standards are based on the integration of different soil evidences of health (physicochemical properties of soil, cropping practices, also soil microorganisms status) into metrics in agro-ecosystems, which can be utilized to assess and steer soil and crop management decision. The cropping system have a major impact on soil health by influencing the way they affect organic matter, the pH level, salinity, contents of nutrients, water dynamics, erosion as well as compaction. To sustain and enhance soil health over time, sustainable agricultural techniques that emphasize biodiversity, organic matter management, and soil conservation foremost are vital. Cropping systems optimize yields, enhance sustainability, strengthen resilience, maximum resource utilization and ensure market access in order to meet the demand of food production. Cropping system has played an essential part in supplying the worldwide demand for food while preserving ecosystems and natural resources for future generation through adopting of cutting-edge technologies, appropriate management methodologies and environmentally friendly techniques.

References

Agrawal, R., Kumar, B., Priyanka, K., Narayan, C., Shukla, K., Sarkar, J., & Anshumali, A. (2016). Micronutrient fractionation in coal mine-affected agricultural soils. *Bulletin of Environmental Contamination and Toxicology, 96*(4), 449–457.

Bainard, L. D., Navarro-Borrell, A., Hamel, C., Braun, K., Hanson, K., & Gan, Y. (2017a). Increasing the frequency of pulses in crop rotation reduces soil fungi diversity and increase the proportion of fungal pathotrophs in a semiarid agro-ecosystem. *Agriculture, Ecosystems & Environment, 240*, 206–214.

Bainard, L. D., Navarro-Borrell, A., Hamel, C., Braun, K., Hanson, K., & Gan, Y. (2017b). Increasing the frequency of pulses in crop rotations reduces soil fungal diversity and increases the proportion of fungal pathotrophs in a semiarid agro-ecosystem. *Agriculture, Ecosystems & Environment, 240*, 206–214.

Barbieri, P., Pellerin, S., Seufert, V., & Nesme, T. (2019). Changes in crop rotation would impact food production in an organically farmed world. *Nature Sustainability, 2*, 378–385.

Basche, A. D., Kasper, T. C., Archontoulis, S. V., Jaynes, D. B., Sauer, T. J., Parkin, T. B., & Miguez, F. E. (2016). Soil water improvements with the long-term use of winter rye cover crop. *Agricultural Water Management, 172*, 40–50.

Baştabak, B., & Koçar, G. (2020). A review of the biogas digestate in agricultural framework. *Journal of Material Cycles and Waste Management, 22*, 1318–1327.

Cercioglu, M., Anderson, S. H., Udawatta, R. P., & Haruna, S. I. (2018). Effects of cover crop and biofuel management on computed tomography-measured pore parameters. *Geoderma, 319*, 80.

Crotty, F., Hannula, E., Hallama, M., & Kandeler, E. (2022). Can soil improving cropping systems reduce the loss of soil biodiversity within agricultural soils? In L. B. Reyes-Sánchez, R. Horn, E. A. C. Costantini (Eds.), *Sustainable soil management as a key to preserving soil biodiversity and stopping its degradation* (pp. 187–220). International Union of Soil Sciences (IUSS)

Dhaliwal, S. S., Sharma, V., Kaur, J., Shukla, A. K., Hossain, A., Abdel-Hafez, S. H., Gaber, A., Sayed, S., & Singh, V. K. (2021). The pedospheric variation of DTPA-extractable Zn, Fe, Mn, Cu and other physico- chemical characteristics in major soil orders in existing land use systems of Punjab, India. *Sustainability, 14*(1), 29.

Dick, R., Myrold, D., & Kerle, E. (1988). Microbial biomass and soil enzyme activities in compacted and rehabilitated skid trail soils. *Soil Science Society of America Journal, 52*, 512–516.

Fierer, N. (2017). Embracing the unknown: Disentangling the complexities of the soil microbiome. *Nature Reviews Microbiology, 15*, 579–590.

Harfield, J. L., Sauer, T. J., & Cruse, R. M. (2017). Chapter 1-Soil: the forgotten piece of the water, food, energy Nexus. In D. I. Spark (Ed.), *Advances in agronomy* (p. 46).

Haruna, S. I., Anderson, S. H., Nkongolo, N. V., & Zaibon, S. (2018). Soil hydraulic properties: Influence of tillage and cover crops. *Pedosphere, 28*, 430–442.

Heather, L. D., Jennifer, M. M., & Manter, D. K. (2024). SEMWISE: A national soil health scoring framework for agricultural systems. *Applied Soil Ecology, 195*, 105273.

Jhariya, M. K., Banerjee, A., Raj, A., Meena, R. S., Khan, N., Kumar, S., & Bargali, S. S. (2021). Species invasion and ecological risk. In M. K. Jhariya, R. S. Meena, A. Banerjee, & S. N. Meena (Eds.), *Natural resources conservation and advances for sustainability*.

Kasper, T. C., & Singer, J. W. (2011). The use of cover crops to manage soil. In J. L. Hartfield, & T. J. Sauer (Eds.), *Soil management: Building a stable base for agriculture* (pp. 321–337).

Khanday, M., Wani, J. A., Ram, D., & Chand, S. (2017). Available nutrients status of soils of forest growing areas of Ganderbal district of Kashmir valley. *International Journal of Chemical Studies, 6*(177–180), 2321–4902.

Kopittke, P. M., Menzies, N. W., Wang, P., McKenna, B. A., & Lombi, E. (2019). Soil and the intensifcation of agriculture for global food security. *Environment International, 132*.

Kumar, S., Meena, R. S., Datta, R., Verma, S. K., Yadav, G. S., Pradhan, G., Molaei, A., Rahman, G. M., & Mashuk, H. A. (2020). Legumes for carbon and nitrogen cycling: an organic approach. In *Carbon and nitrogen cycling in soil* (pp. 337–375). Springer.

LoLorenz, K., & Lal, R. (2018). *Carbon sequestration in agricultural ecosystems*. Springer (1), 978-3-319-92317-8.

Moebius-Clune, B. N. (2016). *Comprehensive assessment of soil health: The Cornell framework manual*. Cornell University

Mudarisov, S., Gainullin, I., Gabitov, I., Hasanov, E., & Farhutdinov, I. (2020). Soil compaction management Reduce soil compaction using a chain-track tractor. *Journal of Terra- Mechanics, 89*, 1–12.

Nouri, A., Lee, J., Yin, X., Tyler, D. D., & Saxton, A. M. (2019). Thirty four years of no tillage and cover crops improve soil quality and increase cotton yield in Alfisols, south eastern USA. *Geoderma, 337*, 998–1008.

Pries, C. E. H., Sulman, B. N., West, C. O., Neill, C., Poppleton, E., Porras, R. C., Castanha, C., Zhu, B., Wiedemeier, D. B., & Torn, M. S. (2018). Root litter decomposition slows with soil depth. *Soil Biology and Biochemistry, 125*, 103–114.

Puissant, J., Villenave, C., Chauvin, C., Plassard, C., Blanchart, E., & Trap, J. (2021). Quantification of the global impact of agricultural practices on soil nematodes: A meta-analysis. *Soil Biology and Biochemistry, 161*, Article 108383.

Schneider, M. K., Luscher, G., Jeanneret, P., Arndorfer, M., Ammari, Y., Bailey, D., Balazs, K., Baldi, A., Choisis, J. P., & Dennis, P. (2014). Gains to species diversity in organically farmed fields are not propagated at the farm level. *Nature Communications, 5*, 41–51.

Shi, R.-Y., Liu, Z.-D., Li, Y., Jiang, T., Xu, M., Li, J.-Y., & Xu, R.-K. (2019). Mechanisms for increasing soil resistance to acidification by long-term manure application. *Soil Tillage Research, 85*, 77–84.

Stavi, I., Bel, G., & Zaady, E. (2016). Soil functions and ecosystem services in conventional, conservation, and integrated agricultural systems, A review. *Agronomy Sustainable Development, 36*, 1–12.

Zuber, S., Behnke, G., Nafziger, E., & Villamil, M. (2018). Carbon and nitrogen content of soil organic matter and microbial biomass under long-term crop rotation and tillage in Illinois, USA. *Agriculture, 8*(3), 37.

Ecological Restoration and Reclamation of Degraded Lands in Indonesia: Eco-Physiological and Anatomy as an Advance Approach

Abban Putri Fiqa, Titut Yulistyarini, Sugeng Budiharta, Fauziah, Trimanto, Lia Hapsari, Apriyono Rahadiantoro, Janis Damaiyani, Sri Een Hartatik, Helbert, Masfiro Lailati, Agusdin Dharma Fefirenta, Mustaid Siregar, Edi Mirmanto, Lindsay F. Banin, and Enny Widyati

Abstract

In this chapter, the use of eco-physiological and anatomical traits of plants for ecological restoration and reclamation of degraded lands in tropical regions is demonstrated. We have mentioned several examples of plants native to Indonesia which can be used to inform land restoration and reclamation in the country. The plants presented in this chapter only resemble a very small portion of large number of Indonesian flora, let alone tropical forests globally which is estimated to contain 40,000–50,000 tree species. Therefore, future efforts are directed to extensively reveal the eco-physiological and anatomical traits of article, various plant species especially trees for ecological restoration. While such assessment will add useful knowledge in restoration ecology, there are several challenges and considerations which might exist as elaborated here.

Keywords

Ecological restoration · Reclamation · Degraded lands · Indonesia · Challenging

A. P. Fiqa · T. Yulistyarini · S. Budiharta · Fauziah · Trimanto · S. E. Hartatik · Helbert · M. Lailati · A. D. Fefirenta · M. Siregar · E. Mirmanto · E. Widyati (✉)
Research Center for Ecology and Ethnobiology, National Research and Innovation Agency of Indonesia (BRIN), Cibinong, Bogor, West Java, Indonesia
e-mail: enny_widyati@yahoo.com

L. Hapsari · A. Rahadiantoro · J. Damaiyani
Research Center for Applied Botany, National Research and Innovation Agency of Indonesia (BRIN), Cibinong, Bogor, West Java, Indonesia

L. F. Banin
UK Centre for Ecology & Hydrology, Penicuik, Midlothian, UK

1 The Status of Forest Management, Degradation, and Restoration in Indonesia

Indonesia has 125 millions ha of forest area in 2023 according to the Ministry of Environment and Forestry. Forest management in Indonesia is conducted by the state government, however, there are complex issues of ecology, economy, and social aspects leading to forest degradation. In ecology view Indonesia covered by large tropical rain forest that rich in biodiversity and home of numerous flora and fauna species. Unfortunately, centralized forest management by the stated is suspected of failing to manage it sustainably including the emergence of illegal logging, change in forest function, forest fire, etc. thereby destroying the country's ecosystem. Many species have lost their natural habitat resulted of deforestation and forest degradation, driving them to the verge of extinction. Forest area also strongly contribute to climate change mitigation due to the vital role of forest in storing carbon dioxide and controlling the planet's climate. In economy sector, the forestry industries have made a significant economic contribution to Indonesia, offering many people employment and mean of income. However, the emphasis on achieving quick financial has frequently resulted in unsustainable logging practices. Significant timber resources has been lost in consequence of illegal logging and unsustainable logging methods, which also decreased the long-term economic potential of Indonesia's forests such as ecotourisms. Sectors relying on healthy forest ecosystem, such as agriculture and tourist, are negatively impacted by deforestation.

There have been numerous issues with land conversion, illegal logging, forest fire, biodiversity loss, deforestation, and disputed over land rights that have raised concerns about

J. A. Parray (ed.), *Soil and Land Use Change*, Sustainable Landscape Planning and Natural Resources Management,
https://doi.org/10.1007/978-3-031-93075-1_9

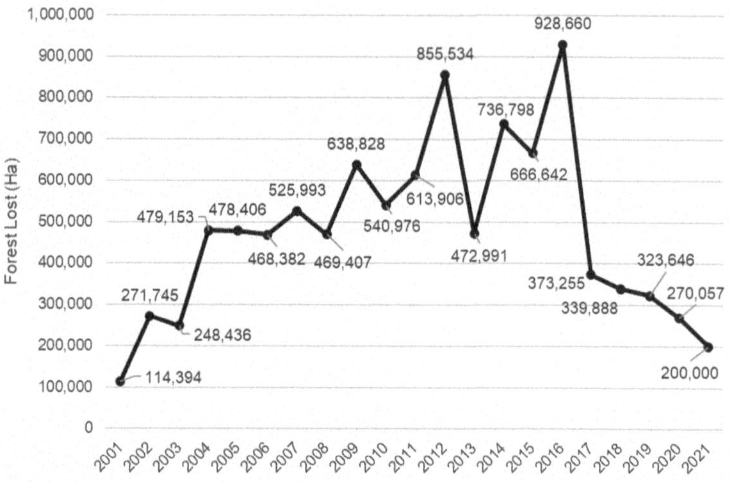

Fig. 1 Indonesia's primary forest loss (2001–2021) (*Source* Mutia, 2022)

current state of forest management in Indonesia. Infrastructure development, logging, and agricultural expansion are the main causes of Indonesia's considerable rate of forest loss, even with efforts to reduce deforestation. It is not debatable, deforestation and forest degradation have negative impacts on biodiversity, carbon emissions, local communities, and ecosystem services. Illegal logging remains a persistent issue in Indonesia, leading to the loss of valuable timber resources and contributing to deforestation. Land conversion for agriculture, particularly for palm oil and pulpwood plantations, has been a major driver of deforestation in Indonesia, leading to habitat loss and environmental degradation. Whereas, forest fires, often linked to land clearance activities like slash-and-burn agriculture, have been a recurring problem in Indonesia, especially during the dry season. Significant volumes of carbon dioxide are released into the atmosphere due to the forest fires, aggravating climate changes in addition to contributing to local air pollution.

It is classified by FAO, the United Nation Organization on Food and Agriculture, in 2022 Indonesia as the third-highest rate of deforestation (0.75 millions ha/year) in the world following Brazil (1.5 millions ha/year), and Congo (1.1 millions hectares/year) (Food and Agriculture Organization of the United Nations, 2020). Based on the Ministry of Environment and Forestry (2022), it shows that the extent of deforestation during 2020–2021 reached 113,534.3 ha, consisted of deforestation in state forest area with 76,236.9 ha (67.1%), and outside state forest areas with 37,297.4 ha (32.9%). Therefore, the Indonesian government implement various policies and initiatives to address deforestation and promote sustainable forest management, such as the moratorium of new licenses on primary forests and peatlands, the One Map Policy, and the establishment of protected areas. Efforts to improve law enforcement, promote sustainable land use practices, and engage with local communities and stakeholders are ongoing to enhance forest management in the country.

Because of various efforts, since at least 2017, the rate of primary forest loss in Indonesia has decreased from year to year (Fig. 1). In 2021, Indonesia succeeded in reducing primary forest loss by 25% compared to 2020. In 2020, Indonesia lost 270 thousand hectares of primary forest land, which is equivalent to 208 metric tonnes of carbon emissions. However, in 2021, primary forest loss decreased to 200 thousand hectares. The success in reducing primary forest loss for five consecutive years is commendable, but forest protection measures still need to be strengthened to maintain this positive trend.

On the other hand, the Ministry of Environment and Forestry stated that there is more than 14 million hectares of degraded lands across the country in 2022. Forest degradation in Indonesia has significant social implications, particularly for indigenous communities and local populations that rely on forests for their livelihoods and cultural practices. Deforestation and land conversion can displace indigenous peoples from their traditional lands, leading to conflicts over land rights and resources. It can also disrupt traditional ways of life and threaten the cultural heritage of these communities.

In Indonesia, land degradation is mostly caused by inappropriate forest management, fires and uncontrolled forest conversion to mining, plantation, industrial, residential and other uses. This is initially caused by a miss-classification of forest based on the functions for production, conservation and protection purposes. While the production and conservation forests are managed by the central government, the protection forests are mandated to the local government which might be unwisely revised and managed according their own interests using the Regional Spatial Plan. In some cases, such revision facilitates forest conversion into other uses. A careless conversion of forest area results in forest and land degradation.

In the last decade, Indonesia has made significant strides in forest restoration, since the government has recognized restoration as a key strategy for mitigating climate change

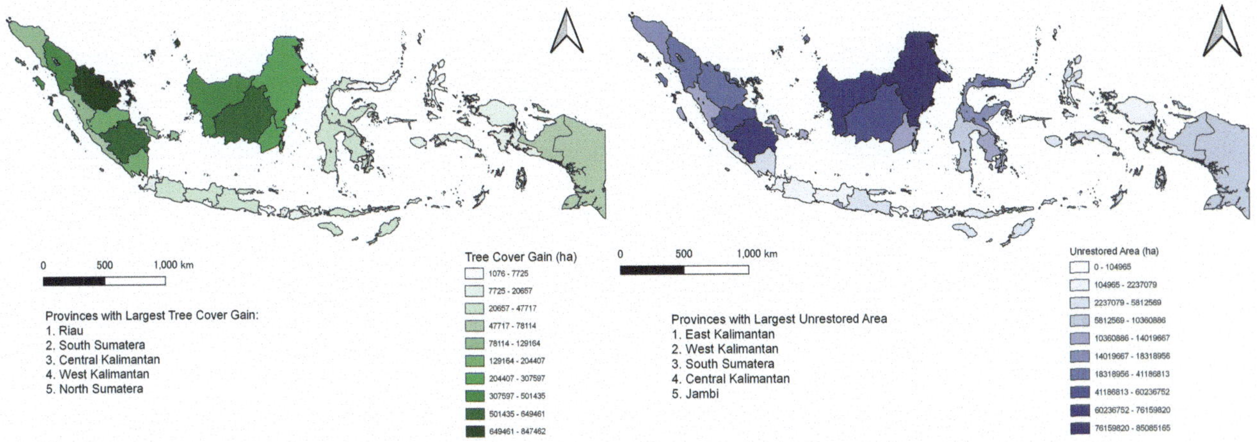

Fig. 2 State of forest degradation and restoration in Indonesia: forest gain (left) and unrestored forest (right). *Source* Global Forest Watch (2024), processed using QGIS 3.22

and improving environmental health. Data from Global Forest Watch in 2024 shows that the degraded forests in Indonesia have proceeded growth with five provinces have the largest gain in their forest area, namely Riau (847,462 ha), South Sumatera (649,461 ha), Central Kalimantan (619,163 ha), West Kalimantan (501,435 ha) and North Sumatra (391,379 ha) (Fig. 2) (Potapov et al., 2022). Nonetheless, a large extent of Indonesia's forest is still unrestored (Fig. 2).

The commitment of Indonesia to degraded land restoration is reflected in the creation of the Peatland and Mangrove Restoration Agency and the Directorate General of Watershed Management and Forest Rehabilitation (Ministry of Forestry of Indonesia, 2022). The State of Indonesia's Forest 2022 toward FoLU Net Sink 2030 by the government (MoEF) provides an overview of Indonesia's forest management strategies, including deforestation reduction and restoration initiatives. To realize the FoLU Net Sink in 2030 the government has an ambitious target that has set a goal of restoring 2 million hectares of peatland and rehabilitating 12 million hectares of degraded land by 2030. The Indonesian Government also launch the Indonesian Ecosystem Restoration Program (PROREKSI), an initiative to restore the function of damaged forest ecosystems. The primary goal of the program is to protect and restore forest ecosytems that have been harmed by a numerous circumstances including land conversion, illegal logging, and forest fires. The main focus is to restore ecosystem function, including maintaining biodiversity, water cycles and ecosystem balance. This program involves a number of initiatives including reforestation, revegetation, fire control, and community awareness and education. Proreksi work at the local to national levels and involves collaboration between government, NGOs and communities.

This demonstrates a strong commitment to large-scale restoration efforts. Those efforts need a combination of stricter policies and restoration initiatives which has led to a significant reduction in deforestation rates in recent years. This creates a more favorable environment for restoration success. Achieving the large-scale restoration goals set by the government will require continued investment, improved coordination between stakeholders, and addressing land tenure issues. Therefore, this paper describes innovations in degraded land restoration emphasizing the characteristics of plants for having optimum adaptation to the degraded lands.

2 Eco-Physiological Plant Traits for Species Selection in Ecological Restoration

The primary objective of ecological restoration and reclamation of degraded land is to enhance biodiversity, improve ecosystem functions, and bolster overall ecosystem health and resilience. Vegetation planting stands out as the principal activity in land restoration and reclamation. Thus, selecting appropriate species is a pivotal step to ensure the success and sustainability of these efforts. In selecting species for ecological restoration, priority is given to those capable of thriving in harsh environmental conditions such as dry climates and nutrient-poor soils prevalent in degraded lands.

Restoration and reclamation efforts require a mix of fast-growing exotic pioneer species and native plants. Consequently, meticulous species selection becomes imperative. Aligning various aspects of planting objectives, species suitability with the location, expected timber and non-timber products, and ecological impacts is crucial. By defining restoration objectives in advance, it becomes easier to choose plant species that can effectively fulfill the desired ecological functions. Considerations of timber and non-timber products to meet community needs and ecological benefits are also factored into species selection. With appropriate plant selection, it is hoped that the survival rate and diversity of planted tree species will increase, thus enhancing ecosystem resilience.

Various methods are employed to select suitable plant species, including their function as nurse or facilitator plants, distribution ability, potential community use, rarity, flagship species, animal feed source, ease of propagation, and resistance to natural disturbances. Increasingly, an approach of utilizing eco-physiological plant traits for selecting plants suitable for restoration and reclamation is gaining prominence. This approach relies on plant characteristics which show the relationship between plants and their environment.

Eco-physiological plant traits encompass morpho-physio-phenological characteristics indirectly influencing fitness through their impact on survival, growth and reproduction (He et al., 2018). These traits, interdependent and often correlated positively or negatively, result from differences in plant biomechanical and physiological needs. Simply put, eco-physiological traits are attributes significantly affecting fitness, survival, and performance. These traits evolve in response to abiotic environmental conditions and interactions with other species. They are pivotal in shaping a plant's performance and can guide plant selection in restoring degraded ecosystems.

Eco-physiological traits can elucidate how plants establish and persist in highly disturbed environments, thereby potentially improving restoration outcomes. This approach aids in understanding how these traits influence ecosystem functions during restoration and reclamation (Zirbel et al., 2017). Each species may have such a combination of response-and-effect traits that matter to the ecological process. These traits serve as primary ecological attributes influencing ecosystem processes and services, explaining resource partitioning and invasive species management (Laughlin, 2014). Understanding these eco-physiological parameters aids in selecting plant species conducive to successful ecological restoration and reclamation efforts. This ecophysiological approach is very useful to be applied to guide plant selection in countries with high biodiversity like Indonesia.

One of the most important groups of plant ecophysiological traits is leaf traits, because these traits are conspicuous in nature and easy to measure in the field in many species. Leaf characteristics have been previously well observed (Liu et al., 2020). It is also stated that the leaf traits are a strongest proxy for whole plant traits compared to stem and root traits. Therefore, adaptation strategies to the environment may be very influential reflected in the leaf traits.

Several leaf trait parameters in plants are classified into different groups. Specifically, leaf morphology are observed in detail including specific leaf area (SLA), leaf mass area (LMA), leaf dry matter content (LDMC), and leaf thickness (LT) (Liu et al., 2020). Stomatal pore length and stomatal density are categorized as stomatal traits. Leaf Nitrogen Content (LNC) and Chlorophyll Content (CC) are considered as biochemical leaf traits. Anatomical traits include cuticle thickness, larger epidermal cells, palisade thickness, and the presence of hypodermis, bundle sheath, extension sclerenchyma, trichomes, and oxalate crystals.

Other leaf traits such as stomatal characteristics, nitrogen and chlorophyll content are also important for indicating the eco-physiological profile of plant species (Xu et al., 2023). As an example of biochemical leaf trait, chlorophyll, associated to LNC, plays a crucial role in photosynthetic rate and is the basic component of the membrane material in plant organs. They are vital indicators of photosynthetic capacity, nutritional status, and growth trends in plants.

The variability of stomata traits demonstrates the different growth strategies on leaf gas exchange and water loss regulation. Stomata serve as crucial regulators, allowing plants to balance the uptake of CO_2 with water loss through transpiration. The size and density of stomata exhibit a trade-off relationship and are intricately linked to various environmental factors, including light intensity, atmospheric CO_2 levels, moisture availability, and temperature, all of which impact gas exchange processes (Bucher et al., 2017). Additionally, anatomical features such as the positioning of guard cells relative to epidermal cells, such as sunken stomata and the deposition of waxes on stomata, can diminish the maximum leaf diffusive conductance (gsmax) (Chaves et al., 2016).

Changes in plant anatomical structure as a response to environmental conditions can occur in various organs such as leaves, stems, and roots. Leaves directly play roles in photosynthetic carbon fixation, respiration, and transpiration (Liu et al., 2020), and among plant organs, leaves are the most plastic in responding to their growing environment, evident from their considerable variation in both morphology and anatomy (Guerra & Scremin-Dias, 2018). Leaves effectively respond to abiotic factors and regulate the balance between photosynthesis and water loss in their growing environment through the regulation of their anatomical structure. Investigating the coordination among anatomical traits can aid in understanding plant adaptation strategies to heterogeneous environments and predicting plant responses to climate change (Guo et al., 2023).

Leaf anatomical traits, including the characteristics of mesophyll, epidermis and vascular tissues can underlie the leaf morphological differences and a series of physiological and biochemical processes of plants (Ni et al., 2022). Modification of stem anatomical structure is a key mechanism through which plants adapt to environmental stress. These modifications in stem anatomical structure enable plants to thrive in diverse environmental conditions by providing added structural support, reducing water loss, storing energy reserves, and enhancing nutrient transport efficiency. Roots are a vital part of plants located underground, serving to provide stability and prevent plants from toppling over. Plant roots play a crucial role as an essential organ for absorbing nutrients, storing, anchorage to substrate, and uptake of water and mineral nutrients. The anatomical structure of fine roots directly reflects their growth and development, with a significant correlation observed between anatomical structure and physiological function (Zhou et al., 2022).

Each plant trait indicates the responses given by plants due to the environmental stress they face. Different stresses encountered by plants are responded to by different traits and can be important indicators in determining the types of plants to be planted in restoration or reclamation areas. Some plant traits (morphological, biochemical, stomatal, and anatomical) commonly observed in various parts of plants, whether in leaves, stems, or roots, indicating their responses to environmental stress are presented in Table 1.

All those plant traits can help plants cope with various stress factors, such as drought, salinity, or nutrient deficiency, by improving their ability to access water, nutrients, and other resources. Understanding these adaptations is crucial for developing strategies to improve crop resilience and productivity in the face of environmental stress.

3 Optimizing Planting Technique

Generally, several steps in preparing plants for planting in restoration of degraded lands are species selection, precondition, and inoculation (Widyati, 2006). Seed source selection of local pioneer species is a critical step, but the most important is seed availability. It is due to the fact that seedlings needed to be planted on the degraded lands are massive, however seeds of local species are generally lacking in seed markets. Next step is seedlings pre-conditioning in the greenhouse. Seedlings can be subjected to controlled stresses, such as drought or limited nutrients, to mimic the harsh conditions they will encounter on the degraded land. This process toughens the seedlings and improves their survival rates. Numerous species of vegetation build symbiotic relationships with mycorrhizal gungi. The symbiosis improve water and nutrients absorption from soil for plants, that is important in degraded lands with poor soil quality. Seedlings can be inoculated with these beneficial fungi before planting.

3.1 Seed Selection

A deep understanding of the physiology of seed germination and seedling establishment is a crucial foundation for the success of reclamation and restoration efforts and can aid in achieving conservation goals and ecosystem recovery. This serves as a key in designing effective strategies for seed treatment, determining the right planting time, and managing irrigation strategies. This knowledge is essential in restoration and reclamation strategies as it plays a role in optimizing environmental conditions, selecting and choosing the right species, resource efficiency, and land management planning.

During seed germination and early seedling growth it is important to optimize the environmental conditions where reclamation or restoration takes place. Environmental factors play a role in the early growth of seeds especially the needs of seed for water, light, and nutrients to assist in designing appropriate strategies to enhance the success of germination and early seedling growth. In this regard, the selection of right species which can optimize environmental conditions is paramount since various plant species have different germination requirements (Donohue et al., 2010). Knowledge of seed germination physiology helps in selecting the appropriate species for specific conditions, such as soil type, climate, and topography. This is important to ensure the sustainability and success of reclamation or restoration projects.

Plants in the early stages of growth also have different efficiency on the utilization of resources such as water, fertilizer, and energy and defense system. Such information is important to develop effective land management strategies including selecting appropriate planting methods, weed control, and protection from pests and diseases (Saturday, 2018). This not only reduces costs but also minimizes unnecessary environmental impacts.

3.1.1 The Physiology of Seed Germination Provides Clues About Seed Treatments

The physiology of seed germination encompasses several processes that occur from the moment the seed is planted until it grows into a seedling. These processes include imbibition, dormancy, germination, and early growth (Penfield, 2017). Each type of plant has different seed characteristics. Based on their storage capacity, seeds are divided into three categories: orthodox, recalcitrant, and intermediate. The following terms are used to describe the physiological characteristics of seeds in response to storage and specific environmental conditions (Walters, 2015).

- Orthodox seeds are seeds capable of enduring dry conditions and can be stored for long periods. Physiologically, orthodox seeds contain low moisture content when mature, enabling them to withstand dry conditions. During storage, orthodox seeds can enter dormancy (a state of rest) to sustain life.
- Recalcitrant seeds are seeds that cannot withstand extreme dry conditions and have limitations in long-term storage capacity. Recalcitrant seeds contain high moisture content when mature and cannot maintain their viability if dried for too long. These seeds are more sensitive to storage and require a more humid environment.

Table 1 Plant traits and their responses to environmental condition related to ecological restoration and reclamation

Traits	Indicator
A. **Leaf**	
a. *Morphological traits*	
Specific leaf area (SLA)	Plant growth rate, productivity, habitat performance
Leaf dry matter content (LDMC)	Plant's resource conservation strategy, carbon storage, plant's ability to withstand drought, plant flammability, leaf life span, resistancy on physical hazard (herbivory, wind, hard rain), soil formation, climate regulation
Leaf mass per area (LMA)	Light absorption, photosynthesis rates, carbon storage, chemical composition, nutrient cycling in the soil, and resistance to herbivores
Leaf thickness (LT)	Greater leaf thickness corresponds to an increased photosynthetic rate, which is adaptive in environments with high light intensity
b. Biochemical traits	
Leaf nitrogen content (LNC)	Photosynthetic rate, litter decomposition rate, leaf life span
Chlorophyll content (CC)	Rate of photosynthetic
c. Stomatal traits	
Stomatal density	Photosynthetic rate, gas exchange, drought tolerancy, water use efficiency
Stomata size	Photosynthetic rate, gas exchange, drought tolerancy, water use efficiency
d. Anatomical traits	
Cuticle tickness	Plant capacity in reflecting light and depleting transpiration
Larger epidermal cells	Reduce water transpiration, prevent sunlight radiation, and reservoir for water in dry conditions
Spongy mesophyll	Resilience of waterlogged and shaded area
Presence of hypodermis	Water reserve to enhance adaptive capabilities in environments with high light intensity
Palisade thickness (number and length of layers)	Water stress tolerant, multi-layered palisade layers indicates adaptivity in high light intensity environments
Presence of bundle sheath extension	Adaptive characteristic for plants under water stress environments
Presence of sclerenchyma	Adaptation strategy to dry conditions
Presence of trichomes	Reduce transpiration rates and lower leaf temperatures, acting as a barrier to water loss and increasing the depth of trapped air layers above the leaf, prevent herbivore and parasite interference
Presence of oxalate crystals	Defense mechanism against herbivores and function as calcium storage in plants, internal source of $CO2$ in plants
B. **Stem**	
a. *Morphological traits*	
Wood density (WD)	Biomass, carbon absorption and storage capacity
Plant height	Plant ability for seed dispersal
b. *Anatomical traits*	
Thickening of cell walls	Under environmental stress like strong winds or mechanical pressure, plants produce thicker cell walls to fortify the stem structure and prevent damage
Increased formation of support tissues	Plants bolster their stems by enhancing the formation of support tissues such as fibers and sclerenchyma. This augmentation enhances stem strength and offers additional support during exposure to mechanical stress
Formation of wax layers	Certain plant species develop wax layers on their stem surfaces to minimize water evaporation and shield the stem from damage caused by dry or hot conditions
Cortex proliferation and xylem thickness	Plants have evolved mechanisms to perceive and respond to mechanical stress, which can involve changes in stem anatomy, including cortex proliferation and xylem thickness, to withstand mechanical stress

(continued)

Table 1 (continued)

Traits	Indicator
C. **Root**	
a. Morphological traits	
Specific root length (SRL)	Ability to utilize limited resource
Root N content (RNC)	Ability of N fixation
Root mass	Water uptake and nutrient uptake
b. Anatomical traits	
Cortex proliferation	Thicken in response to stress, providing additional support and protection to the root
Xylem thickness	Plant's ability to transport water and nutrients
Root hair development	Root hairs can be affected by stress, leading to changes in their length and density, which can impact nutrient uptake
Root elongation	Reduced elongation in response to drought or other stressors
Root branching	Producing more lateral roots to increase their ability to absorb water and nutrients

- Intermediate seeds have moderate moisture content when mature. Intermediate seeds often have greater flexibility in storage and distribution compared to recalcitrant seeds, but may not be able to be stored for very long periods like orthodox seeds (generally less than 1 year).

Understanding the factors influencing seed germination such as temperature, humidity, and planting depth is crucial to treat the seeds most effectively, enhancing the germination rate and seedling growth. Each plant species has specific requirements for optimal germination. For example, some seeds require special treatments like stratification or scarification to break dormancy and stimulate germination. Orthodox seeds typically have hard seed coats, requiring treatment to break seed dormancy. The advantage of selecting orthodox seeds lies in their ability to withstand extreme environmental conditions such as drought, soil nutrient deficiency, or extreme temperatures. This makes them suitable for planting in degraded lands. Orthodox seeds can endure drying at low moisture levels (Solberg et al., 2020). For instance, legume seeds are considered orthodox and generally exhibit physical dormancy. Some endemic plants have recalcitrant seed characteristics, including species members of Shorea which cannot be stored for long periods. The majority of dry tropical species have orthodox seeds with dormancy traits, which can be overcome by mechanical scarification, hot water treatment, and acid scarification.

3.1.2 Seed Physiology Determines Optimal Time for Planting

Seed characteristics determine the best time for planting seeds in restoring and reclaiming degraded lands. Each species has an optimal period for germination and early growth, so that the appropriate planting time of seeds can be determined to have the best chance for the seedlings to grow and thrive. Seedling formation involves a series of processes aimed at producing high-quality seedlings. This process begins after the seed has germinated and includes the early growth phase of the plant, including root formation, leaf development, and adaptation to the new environment.

Each plant species has different physiological requirements to initiate the germination process which are influenced by temperature, humidity, and light which can be used to determine the optimal time for planting seeds to increase the chances of success. Some seeds require specific cooling (cold stratification) while others needs heating/dry periods before they can germinate effectively. The availability of light, temperature, and nitrogen are the most important environmental factors that have a significant impact on seed dormancy and germination. Without a good understanding of germination physiology, it might result in the wrong planting time, causing failure in land restoration efforts.

The right planting time contributes to increasing the survival rate of seedlings. Seedlings planted at the appropriate physiological stage will have a better chance of rooting well and surviving in the new environment. Understanding seed physiology allows for better assessment of environmental conditions that may affect early plant growth. For instance, if specific seeds are sensitive to drought during the early germination stage, then selecting a planting time more suitable for the rainy season or providing additional irrigation can mitigate risks.

3.1.3 Seed Characteristics Related to Water Needs and Irrigation Strategies

The preparation of seedlings for the restoration and reclamation of degraded land involves a series of processes aimed at producing high-quality seedlings. Knowledge on the physiology of seedling formation is essential to select quality planting materials, ensure the use of appropriate planting media, and provide proper care to produce superior seedlings. Water fulfillment and efficient irrigation management is crucial in maintaining the growth and survival of seedlings. The early stages of plant growth after germination require special attention to water needs and irrigation to support healthy growth. Understanding seed physiology aids in planning irrigation strategies that prevent excessive water loss and maintain the appropriate moisture for the germination process.

Environmental factors greatly influence water availability for plants and should be considered when designing appropriate irrigation strategies for specific environmental conditions. Water availability and irrigation intensity should take into account the needs of each species since each type of seed has different water requirements to trigger the germination process. Some seeds have hard or impermeable outer layers that require special treatments to allow water penetration and trigger germination (Pedrini et al., 2020). Irrigation strategies should consider the need for pre-germination treatments to ensure that seeds can efficiently absorb water.

Growing conditions such as temperature, humidity, and water flow rate around the seeds will affect the seeds' ability to absorb water. Irrigation strategies should be adjusted according to environmental conditions that are suitable for seed physiology. On the other hand, if seeds are exposed to water deficiency for too long, embryo death may occur before germination. Understanding seed physiology helps plan irrigation strategies that prevent excessive water loss and maintain the proper moisture for the germination process. However, inappropriate humidity levels can lead to the growth of fungi and other pathogens that can damage seeds and inhibit germination (Magan et al., 2004). Therefore, irrigation strategies should consider this risk and take measures to control it.

The importance of understanding seed physiology also relates to determining the planting media. Planting media variations can impact biomass, morphological characteristics, and seedling growth (Chirino et al., 2011). In the context of restoration in drylands, a significant challenge in forest restoration is to improve the water-holding capacity of the planting media used in nurseries. This will allow seedlings to maintain adequate hydration, reduce post-transplantation stress, and improve water status during the first months after planting.

3.2 Root Anatomy Can Guide Planting Depths and Spacing for Optimal Water and Nutrient Acquisition

The root of plants are the first part that grow from seed in germination process. The development and differentiation of the upper portion to form shoots comes next. Roots play key roles in anchoring plants into soil, acquisitioning water, and adapting to water-limited environments. Schenk and Jackson (2002) explained power of root penetration and phenology define plant's rooting depth (RD). The older plant growing, generally, the deeper it roots. The study also reported that RD also strongly determined by soil type, solum thickness, and belowground resources availability. Rooting dept reflects the maximum depth reached by root affect the volume of exploration area of plants in soil.

Root anatomy and hydraulic conductivity are essential factors in the uptake of soil water, particularly water deep in the soil. Root anatomical features are significant tools for crop selection that maximizes the use of deep soil water. A better understanding of root phenotypes associated with deep root growth and activity would benefit in the development and breeding of perennial crops for improved subsurface exploitation (Clement et al., 2022). Root anatomy phenotypes can improve adaptation to abiotic stresses such as low nitrogen and phosphorus (Lynch, 2013), water deficiency or flooding (Jaramillo et al., 2013).

Root anatomy has a significant impact on tissue composition as well as the fraction of metabolically active roots. In most plant species, the root cortex's parenchyma cells contain the most protoplasm in primary root tissue. Other metabolically active cells in the epidermis, endodermis, and living cells of the xylem and phloem make up a small proportion of root volume, except in highly aerenchymatous plants like rice, where the stelepith, epidermis, and endodermis contain proportionally more protoplasm (Jaramillo et al., 2013).

There is growing evidence that by lowering the metabolic expenditures associated with soil exploration, root cortical aerenchyma (RCA) can enhance the uptake of water and nutrients. While most studies on respiration-induced changes (RCA) focus on the effects of hypoxia (Jackson & Armstrong, 1999), RCA can also be brought on by a variety of abiotic stressors in normoxic environments, including as drought, high temperatures, and low levels of N, P, and sulfur (S) (Zhu et al., 2010). The experiments offer compelling proof that by lowering the metabolic cost of soil exploration, RCA enhances the acquisition of soil resources. For roots to penetrate hard soils, root hairs are just as important as root diameter. In mechanically hindered soil, root hairs enhance root penetration.

3.3 Seedling Improvement

Seedling improvement is also a crucial technique in degraded land restoration since the planted seedlings need to be well-adapted to the harsh conditions. These restoration plants should be fast-growing, drought-tolerant, and nitrogen-fixing (if possible) to improve soil fertility. Seedlings can be engineered by inoculation with beneficial microbes such as mycorrhiza. Mycorrhizal fungi has an important rule in promoting seedling establishment and restoration success on degraded lands (Khaliq et al., 2022).

Mycorrhiza enhances nutrient uptake. Almost all degraded lands are deficient in essential nutrients like phosphorus, nitrogen, and potassium. Mycorrhizal fungi form a symbiotic relationship with plant roots, extending their reach and increasing their efficiency in absorbing these vital nutrients from the soil. This improved nutrient uptake allows seedlings to grow stronger and healthier, especially in nutrient-poor environments. Mycorrhiza increases the capacity of root surface to absorb more nutrients uptakes. They build a mutual association with the roots that aids in the absorption of numerous minerals such as P, N, and Ca (Prasad et al., 2017).

Mycorrhiza helps plant increased stress tolerance. Seedlings on degraded lands face various stressors like drought, nutrients deficiency, and heavy metals. Mycorrhizal fungi can help plants cope with these stresses by producing hormones and compounds that enhance tolerance. This allows seedlings to establish themselves and thrive in harsh environments (Bahadur et al., 2019).

Association between plant and mycorrhiza improves water acquisition. Degraded lands often suffer from water scarcity due to factors like erosion and reduced organic matter. Mycorrhizal fungi facilitate plants access higher amount of water from a wider region of the soil by functioning as the root system extension. When growing in drought stress plants release strigolactone, a chemical signaling to arbuscular mychorrhiza (AMF) (Khaliq et al., 2022). After that, AMF settles on the root plants and aids in their ability to collect water from a wider surface area. This improved water acquisition is crucial for seedling survival during dry spells and drought conditions (Bahadur et al., 2019).

Mycorrhiza association in plant roots improves soil structure by binding soil particles together with a network of fungal hyphae. AMF gives beneficial alterations in the soil structure around plant roots hence improving soil biological, physical, and chemical characteristics. This creates better aeration and drainage within the soil, leading to healthier root development and increased water infiltration (Gutjahr & Paszkowski, 2013). This improved soil structure benefits not only the mycorrhizal seedlings but also the overall health of the restored ecosystem.

Many previous studies reported that plants discovered in degraded lands are prevented from poison of heavy metals when mycorrhiza are established in their roots (Galli et al., 1994). There are several ways how mychorrhiza protect roots, through metal immobilization and heavy metals localization within the mycelium and mycorrhizal roots, metal concentration as crystalloids and save it within both the mycelium and cortical cells of mycorrhizal roots (Galli et al., 1994). A species of AMF *Gigaspora margarita* is one of general symbiont which is assemblage association with several plant species, result in plant toleration to soil contaminated by As, Hg, Pb, Al, and Cu (Rollon et al., 2017).

4 Enhancing Plant Survival and Growth

Enhancing degraded land restoration can be achieved by both improving underground (soil amendment) and aboveground (seedling adaptability). Soil amendments, generally by improving soil organic matter, create a more favorable environment for plant establishment and growth. This kick-starts the restoration process, allowing vegetation to take root and gradually improve the overall health of the degraded land.

4.1 Soil Amendment

Soil amendment is conducted for improving soil structure and soil fertility, and boosting soil microbiome. As well as for enhancing water retention and reducing soil erosion. Amend soil with organic matter improves soil fertility, it is due to degraded land often suffers from nutrient deficiencies, reduced organic matter content, and poor soil structure. Amendments like compost, manure, or biochar can replenish essential nutrients like nitrogen, phosphorus, and potassium. This creates a more fertile environment for plant growth, promoting the establishment of vegetation and kickstarting the restoration process (Gunarathne et al., 2017).

Amending soil with soil organic matter also enhances water retention. It is not debatable that many degraded lands struggle with water scarcity. Amendments like organic matter or certain clays can improve soil moisture retention. This allows plants to access water for longer periods, especially during dry spells, and reduces the risk of drought stress.

Soil amendment also boosts soil microbiome, because plants need healthy soil teeming with beneficial bacteria, fungi, and other microorganisms. These organisms play a critical role in nutrient cycling, decomposition, and plant health. Amendments like compost or composted manure can introduce these beneficial microbes back into degraded soil, fostering a healthy soil ecosystem that supports plant growth (Wang et al., 2022).

The increased soil organic matter content also improves soil structure. Generally, degraded land often has compacted soil, hindering water infiltration and root development. Organic matter amendments can improve soil structure by creating larger pore spaces. This allows for better air and water movement within the soil, leading to healthier root growth and improved overall soil health (Priori et al., 2021). It also reduces erosion, since bare, degraded land is susceptible to erosion from wind and water. Amendments like cover crops or mulch can help protect the soil surface, reducing erosion and preserving valuable topsoil.

Widyati (2006) reported, soil amendment with sludge of paper mills (50% v/v) improved soil pH of ex-coal mining land from 2.8 to 6.6 in 14 days after implementation. The treatment also increased macro nutrients content such as nitrogen, phosphates, potassium, as well as reduced the excessive micronutrients Fe, Mn, Zn, Cu, significantly. The most important, sludge also reduces sulfate content that is highly toxic in the environment.

4.2 Plant Physiological Assessment

The critical point of restoration and reclamation of degraded lands is assessment of the treatment. Physiological assessments are crucial tools for understanding how plants respond to various environmental factors (including the treatment). The intervention when restoration and reclamation (i.e.: soil amendment, pH adjustment, fertilizers, etc.) could be reflected on plant physiological changes. Here are some plant physiological assessment techniques that has been widely used:

4.2.1 Analysis of Chlorophyll Fluorescence

Chlorophyll molecules in plant leaves can re-emit light when returning from excited state to non-excited states. This light can be used as an indicator of energy conversion of the photosynthesis process. By measuring the fluorescence light we can estimate the plant photosynthetic activity (Murchie & Lawson, 2013). This method is a non-destructive mode.

Analysis of chlorophyll fluorescence (CF) could be used to measure most types of environmental induced plant stresses (e.g., temperature, light, water, metals, etc.). When stressed, plant metabolism is disturbed and can be reflected as an imbalance of light energy absorption by chlorophyll and the energy spent in photosynthesis (Murchie & Lawson, 2013). For the CF measurement, handheld measurement apparatus are commonly used. The comparison of the pre-photosynthetic fluorescent state (dark-adapted) or F0 (minimum fluorescence), to maximum fluorescence (Fm) where saturating light activates the maximum number of reaction centers in Photosystem II is used to calculate Fv (variable fluorescence). The ratio of Fv to Fm is called maximum photochemical efficiency (Fv/Fm). In general, when plants are under greater stress, the number of available open reaction centers in Photosystem II is decreased, resulting in the lowered value of Fv/Fm (Murchie & Lawson, 2013).

4.2.2 Antioxidant Activities

Normally, all organisms are living in a balanced redox state. When disturbances occur, the production of peroxides and free radicals could damage all cell components (DNA, Proteins, Lipids). The imbalance of the reactive oxygen species/ROS (i.e., O_2-, –OH and H_2O_2) presence in the plant system, and the ability to detoxify the reactive intermediates or repair the damage is called oxidative stress (Ramakrishnan et al., 2022).

Plants have numerous activities of antioxidant functioning as indicators, such as the level of superoxide dismutase (SOD), catalase (CAT), and malondialdehyde (MDA). The SOD is an acting as an antioxidant to shield cells from ROS induce oxidative damage. The catalyzations are as follows: 2 molecules of O_2- are reacted with 2 molecules of proton donor (H+) and converted to two less damaging species (O_2 and H_2O_2). In plants, these SOD are coordinated by metal coenzymes (Fe, Mn and Cu) (Alscher et al., 2002).

Catalase (CAT) is an enzyme commonly found in all organisms that are exposed to oxygen. CAT is degrading enzyme responsible for breaking hydrogen peroxide (H_2O_2) out water (H_2O) and oxygen (O_2) (Sharma & Ahmad, 2014). The CAT activity could be measured by detecting the degradation of H_2O_2 or by O_2 release. A UV spectrophotometer can be used for this measurement, where H_2O decrements per unit time at 240 nm. The CAT activity is defined as the number of enzymes needed to change the absorbance of 0.001/min under assay conditions.

Malondialdehyde (MDA) is an end product of degradation of polyunsaturated lipids by ROS. MDA content can be measured by reacting samples with thiobarbituric acid, resulting in a fluorescent red derivative that can be detected and measured by spectrophotometer (Zhang et al., 2021).

4.3 Monitoring and Evaluation Using Physiological and Anatomical Markers to Track Plant Health and Restoration Progress

The results of restoration may not be immediately apparent. Thus far, the success of restoration has often been judged solely by the extent of vegetation cover, overlooking other essential aspects associated with the goals of the activity. The criteria for evaluating restoration success have not been standardized as this activity is complex and multifaceted. Common parameters used to assess restoration success include the vegetation formed, species diversity, and the ecosystem services generated (Suding, 2011).

The simplest assessment method involves observing the structure and composition of the vegetation post-restoration. A well-structured and composed vegetation provides suitable habitat for local organisms, maintains ecosystem productivity, and aids in predicting succession trends in the revegetated area. Vegetation plays a crucial role in soil and water conservation, providing litter to sustain the organic matter cycle in the soil. Its canopy protects soil from raindrop impact, preventing soil aggregate damage and erosion, while stem flow during rainfall helps increase surface water availability (Puigdefabregas, 2005).

Species diversity is also utilized as a measure of revegetation success. Heterogeneity in an area promotes greater ecosystem stability compared to homogeneous areas. The more diverse the species in an ecosystem, the greater the ecological services it can provide (Wu, 2013). Increased vegetation in an area fosters diverse organisms utilizing it as habitat or food source. Indigenous species growth indirectly conserves local fauna.

As noted by Clewell and Aronson (2013), ecosystems are dynamic and evolve in response to natural changes and human activities, with species composition and quantity regulated by internal processes. Like species evolution, ecosystem evolution can occur gradually or abruptly. Habitat management to support restored species may require long-term, even indefinite, commitment in some cases. While restoration activities are often monitored for only a few years, true evaluation may require decades to determine whether planted species can sustainably thrive and reproduce. Hence, restoration is a long-term commitment necessitating appropriate monitoring planning aligned with the life cycles of planted species. Through careful monitoring, comprehensive insights into future population growth can be attained.

The survival of plant growth within restored areas is the most crucial aspect of restoration success itself. When planted vegetation thrives and initiates the emergence of natural succession on track, it undoubtedly represents the desired success in the restoration process. One way to understand how plants can adapt to new environments with environmental stress is by monitoring plant growth based on their functional traits.

The adaptive capacity shown by plants in restoration areas, one way to understand how a plant can adapt in restoration areas is by comparing its functional traits with the same species in reference areas. The reference area becomes one of the important variables in restoration activities. The presence of a reference area will facilitate determining the goals of restoration activities. The reference area becomes a model to be achieved from restoration activities. Furthermore, the presence of a reference area will also help in monitoring and evaluating whether this activity has been carried out under the expected end goals (Choi, 2004).

The functional ecology approach, especially the use of plant traits, can be a highly useful tool in evaluating ecosystem restoration success. Functional trait features show an individual's level of adaptation as well as their methods for using resources from and to adjacent trophic levels. Moreover, an ecological niche is defined by all of a species' functional characteristics. Therefore, the worth and variety of species as well as their traits that affect how communities and ecosystems function are represented by functional diversity (Gastauer et al., 2018). Within a functional ecosystem, functional ecology can aid in understanding the functional characteristics of planted plants in restoration projects, thereby assisting in evaluating how well the restored ecosystem can provide necessary ecosystem services such as habitat provision for wildlife, soil erosion control, or water filtration (Carlucci et al., 2020; Laughlin, 2014).

Regarding resistance and resilience, plant traits related to resistance and resilience to environmental pressures can be measured, such as drought or fire resistance and the presence of invasive species, providing insights into how well the restoration ecosystem can withstand or recover from external disturbances. Success aspects in plant selection can be determined through the analysis of functional traits such as seed size, growth rate, or herbivore resistance, thus evaluating the success of plant selection planted in the ecosystem to be restored. Plants with certain characteristics may be more likely to survive and thrive in new environments. Community interactions can also be determined, understanding how plant traits interact with other species in their community, both positively (e.g., through pollination) and negatively (e.g., through competition for resources), providing insights into the structure and function of the restored community. Furthermore, evaluations of resilience to environmental changes can be known from certain observable characteristics indicating the adaptive capacity to environmental changes caused by climate change or human activities. Selecting plants with suitable traits can help enhance the resilience of restoration ecosystems to unforeseen environmental changes (Carlucci et al., 2020).

By employing a plant trait approach in restoration evaluation, we can gain a more comprehensive understanding of the success of restoration projects and design more effective management actions to support the recovery of healthy and sustainable ecosystems. Our study conducted to evaluate the success of post-mining restoration in the concession area of PT. Trubaindo Coal Mining (PT. TCM) in East Kalimantan using a leaf trait approach yielded very interesting results.

4.3.1 Morphological Trait

As a morphological trait, leaf size or Leaf Area (LA), can affect a plant's ability to absorb light, photosynthesize, and transpire. These characteristics are important in evaluating water and energy use efficiency and play a significant role in soil conservation. Wide leaf size in the upper stratum can collect rainwater, posing a risk if it directly falls to the ground. The presence of plants in lower strata, with smaller leaf area sizes, can help break the force of raindrops falling to the ground, thus preventing soil structure damage. Leaf area is one trait with high heritability values, indicating its ability to adapt to environmental changes. Leaf size is more controlled by genetics and inherited variations, making it potentially more adaptable through natural selection, even compared to SLA (Ahrens et al., 2020). The study on some species in the restoration area showed that plant species planted in dry and high light exposure tend to develop narrower leaf area, than species planted near a water resource.

Another morphological trait, leaf thickness (LT), reflects the ratio of leaf volume to leaf area, determined by the density of mesophyll cells per unit area (Nardini, 2022). It is a quantitative trait influenced by environmental factors such as water availability, temperature, and light intensity (Coneva et al., 2017). Based on leaf thickness analysis, species planted on the restoration area of PT. TCM, *Robroshorea balangeran* had the thickest leaves, while *Litsea artocarpifolia* had the thinnest leaves. Globally, plants in dry regions tend to have thicker leaves as an adaptation to enhance their ability to retain water when facing low water availability (Poorter et al., 2009). At the cellular level, leaf thickness is generally associated with increased cell size, which can support water storage, and is evolutionarily advantageous for efficient water use and reduced transpiration (Guo et al., 2017). Research findings indicate that leaf thickness may increase in response to high light intensity and high temperatures (Feng et al., 2019). Water scarcity, such as during drought conditions, significantly affects leaf epidermis and mesophyll tissue thickness, resulting in thicker leaves (Klich, 2000).

Leaf area-to-mass ratio, indicated by SLA, LDMC, and LMA, which is another example of leaf morphological traits, can indicate plant adaptation to specific environmental conditions. Together with Leaf Area (LA), SLA, LDMC, and LMA are known morphological trait. Leaves with high SLA values can indicate their ability to maximize resources within the ecosystem. Plants with high SLA values also demonstrate their ability to compete with other species. Therefore, in restoration objectives aimed at restoring the original forest structure, local plants with high SLA values are predicted to compete with invasive species threatening to occupy open areas, reducing the risk of invasive species spreading in the restoration area (Funk et al., 2008).

Our study conducted in the reclamation area of ex-coal mining of PT. TCM in East Kalimantan revealed that species such as *Alstonia angustifolia*, *Dyera costulata*, and *Syzygium myrtifolium* found both in reference areas, specifically in natural forests and those planted in reclamation areas, experienced significant differences. *A. angustifolia* and *D. costulata* showed a similar phenomenon, where the SLA values of plants in natural forests were higher than their LDMC values. However, when planted in reclamation areas, the SLA values of plants significantly decreased, while the LDMC values tended to increase beyond their SLA values. In contrast to the two previous species, *S. myrtifolium* exhibited a different phenomenon where the SLA value in the reclamation area increased, while its LDMC value decreases compared to the natural forest (Fig. 3). The LDMC and SLA values of the natural forest area for the species *A. angustifolia* and *D. costulata* were located in Melak, East Kalimantan (Fiqa et al., 2024), while for *S. myrtifolia,* samples from the natural forest

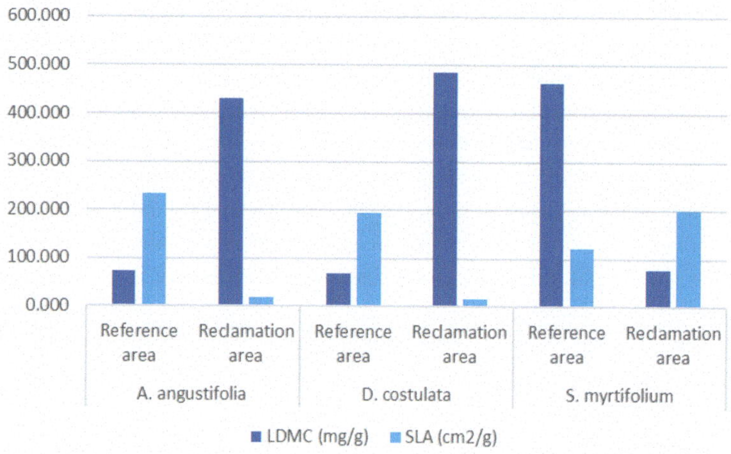

Fig. 3 SLA and LDMC value of species found in both reclamation and natural habitat in the concession area of PT. Trubaindo Coal Mining in East Kalimantan (*Source* Yulistyarini et al., 2024)

were taken from the Prigen area, East Java (Hapsari, 2022). LDMC indicates the ability of plants to survive in dry areas, so when plants can adapt by increasing their LDMC value, they will survive better in reclamation areas that tend to be dry and exposed to high sunlight. Meanwhile, species that maintain high SLA values will be less able to survive in dry reclamation areas. These three species can serve as examples of which species can grow well and withstand in reclamation areas, and which cannot.

Our study also shows a different result compared to previous research, where at the age of 14 years of reclamation, certain species' SLA values have already changed in their efforts to adapt to the environment. The SLA is a trait that tends to be conservative and less evolutionary plastic because it is a major structural trait that limits leaf shape variation in plant taxonomy. Poorter et al. (2009) mentioned that SLA is more determined phylogenetically than environmental factors, although there is still quite a large intraspecific variation. Furthermore, the newest theory based on a global analysis by Siefert et al. (2015) of 6,714 species indicates that SLA is one of the most evolutionarily conservative traits and less responsive to environmental variations compared to other traits in the leaf economics spectrum. More studies are needed, especially in tropical regions where research is still very limited, to show that SLA and LDMC values can also be used as indicators of the success of species adaptation planted in reclamation areas.

4.3.2 Stomatal Trait

In our study in the concession area of PT. TCM, species growing on reclamation areas, characterized by higher sunlight exposure, limited water availability, acidic soil pH, and low soil nutrients, tended to have smaller stomata and more papillated epidermal cells arranged densely, covering their stomata (Fig. 4c and d) than in natural habitat (Fig. 4a and b). This structure provides additional protection against detrimental environmental factors such as UV radiation, drought, and pathogen attacks (Jordan et al., 2005) (Fig. 4e). The reduction in stomatal size in drier areas is a form of plant adaptation (Haworth et al., 2023). Smaller stomata can promote greater photosynthetic rates, particularly under fluctuating environmental conditions (Harrison et al., 2020). Plants with large and few stomata tend to have higher water-use efficiency, but also reduced rates of photosynthesis, compared with plants that have many but smaller stomata (Kardiman & Ræbild, 2017). A slow opening and closure of stomata could lead to sub-optimization of gas exchange and loss of water through transpiration as well (Kardiman & Ræbild, 2017).

4.3.3 Anatomical Traits

The results of our study on several species planted on post-mining reclamation area of PT. TCM in East Kalimantan showed variations in anatomical structures among species (Yulistyarini et al., 2024/TCM report). Stresses are highest

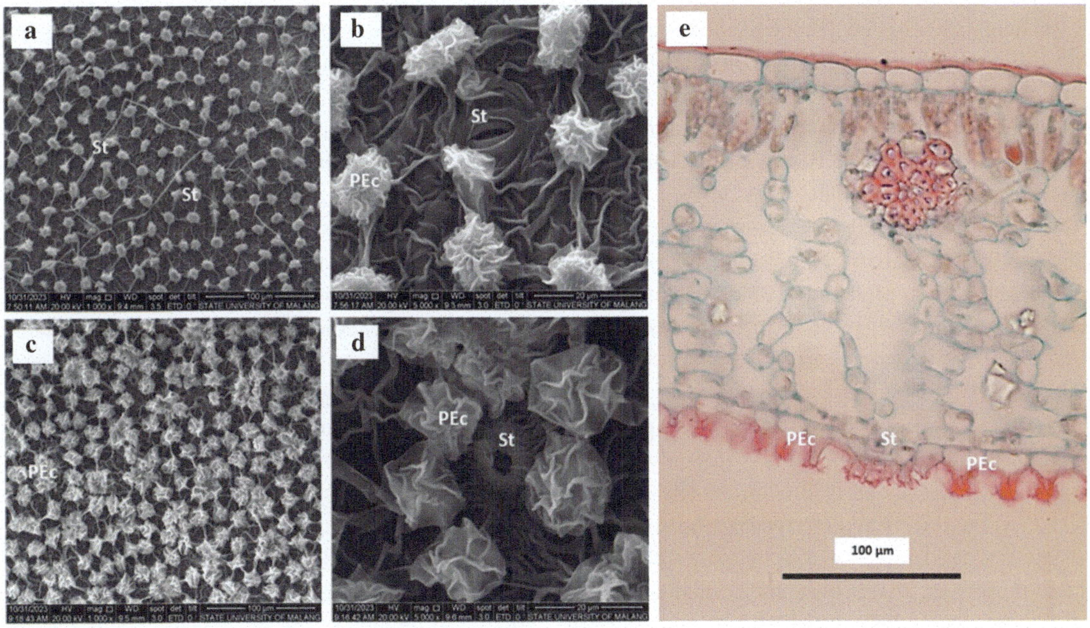

Fig. 4 Stomatal appearance of *Dyera costulata* in natural forest and planted in reclamation area of PT. Trubaindo Coal Mining in East Kalimantan. Ultrastructure of the epidermal surface of *D. costulata* in natural forest area (SEM, magnified 1000 × and 5000x) **a**, **b**, and in reclamation area (SEM, magnified 1000 × and 5000x) **c**, **d**, cross-section of *D. costulata* leaf with papillated epidermis cells structure present on the abaxial surface (LM, magnified 400x) (E). St: stomata, PEc: papillated epidermis cells (*Source* Yulistyarini et al., 2024)

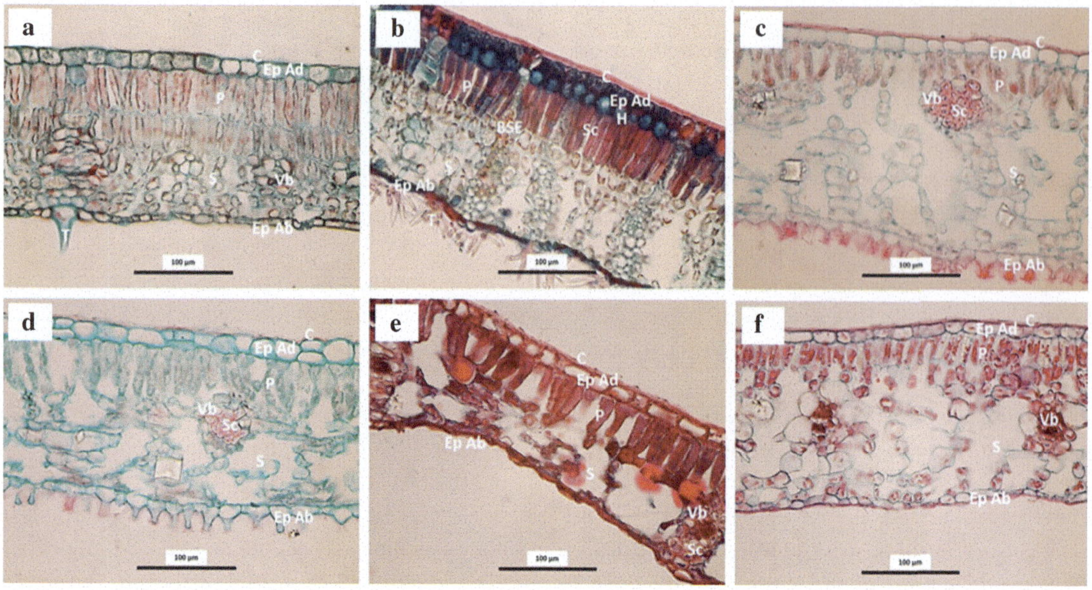

Fig. 5 Anatomy structures of six plant species planted on post-mining reclamation area of PT. Trubaindo Coal Mining in East Kalimantan: *Neolamarckia cadamba* **a**, *Rubroshorea balangeran* **b**, *Dyera costulata* **c**, *Alstonia angustiloba* **d**, *Litsea artocarpifolia* **e**, *Syzygium myrtifolium* **f**. C: Cuticle, Ep Ad: Adaxial Epidermis, Ep Ab: Abaxial Epidermis, H: Hypodermis, P: Palisade Mesophyll, S: Spongy Mesophyll, Vb: Vascular Bundle, Sc: Sclerenchyma, T: Trichome, BSE: Bundle Sheath Extension (Yulistyarini et al., 2024)

on the surface of a body, therefore the cuticle deposited on the outer surface of organs would play an important role as a structural element adding mechanical support to the epidermal tissue (Dominguez et al., 2011). Cuticles prevent uncontrolled diffusion of water into the atmosphere (Goodwin & Jenks, 2005). All leaves have a cuticle that protects the leaf and helps to keep water inside. When the cuticle is thick, the plant keeps even more water inside. Thick cuticles are an adaptation to help the plant survive in hot and dry environments. The results of our study from the reclamation area indicated that *R. balangeran* had the thickest cuticle layer, while *S. myrtifolium* had the thinnest cuticle (Fig. 5b and 6f). In this study, the cuticle of *R. balangeran* had a thickness of 3.42 ± 0.25 µm, while that of *S. myrtifolium* was only 1.09 ± 0.13 µm. In plants, cuticle varied > 100 times across species from < 0.05 to > 10 µm (Jeffree, 2006).

R. balangeran is known to have the thickest epidermal tissue compared to other species, and it is the only species that possesses a hypodermis derived from epidermal tissue (Fig. 5b). The hypodermis is a layer beneath the epidermis. Together with the epidermis, this layer plays a role in shielding the mesophyll tissue beneath it from excessive sunlight radiation, particularly ultraviolet A and B, and lowering the leaf's internal heat load (Roth, 1984). Rindyastuti et al. (2018) also explained in their study on Sempu Island that the presence of the hypodermis in *Dendrobium subulatum* serves as a crucial structure for water retention and as an adaptation to habitats with high irradiance to reduce leaf transpiration.

The mesophyll is a crucial structure in leaves composed of palisade and spongy tissues, both playing essential roles in photosynthesis by housing chloroplasts where the photosynthetic reaction occurs. Plants with thicker mesophyll layers likely have better photosynthetic capacities (Liu et al., 2020). The spongy tissue facilitates gas transport due to its larger intercellular spaces, while the palisade tissue aids in enhancing leaf photosynthesis. In our study, among the six species, *N. cadamba* had the thickest palisade tissue, compact and consisting of two layers, followed by *R. balangeran* (Fig. 5a and b). In contrast, *S. myrtifolium* had the thinnest palisade tissue, comprising only one layer (Fig. 5f). Other studies indicate that plants resistant to water stress tend to have thicker palisade cells (Poejiastuti, 1994). Palisade cells enlarge and increase chloroplasts in water-stressed environments (Poejiastuti, 1994) and its thickness correlates with greater leaf tolerance to water stress. In environments with higher light intensity, palisade cells tend to be longer and have more than two layers (Taiz & Zeiger, 2002).

According to Lack and Evans (2001), the vascular bundle is a transport system that includes xylem and phloem that are closely related to the movement of nutrients and water. The arrangement of vascular bundles in the six species varies widely, with only two species undergoing Bundle Sheath Extension (BSE), namely *R. balangeran* (Fig. 5b) and *L. artocarpifolia*. BSE, which connects leaf vascular bundles to

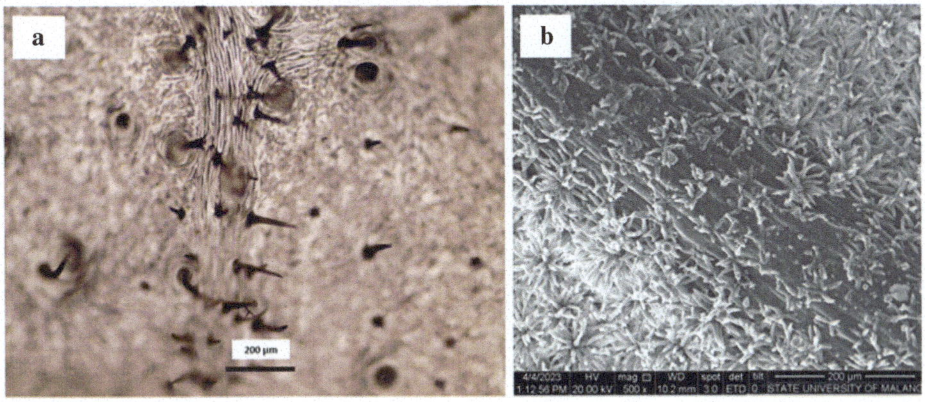

Fig. 6 Different types of leaf hairs (trichomes) in the abaxial surface. The unicellular trichomes type in *N. cadamba* (LM, magnified 100 ×) **a**, the stellate (star-like) trichomes type in *R. balangeran* (SEM, magnified 500 ×) (*Source* Yulistyarini et al., 2024)

the adaxial surface epidermis layer, is an adaptive characteristic for plants in water-stressed environments as it functions to transport water to vascular bundles, epidermis, and photosynthetic tissue (Evert, 2006). Additionally, the presence of sclerenchyma in the vascular bundle and BSE as observed in *R. balangeran* and *D. costulata* (Fig. 5b and c) also indicates plant adaptation strategies to dry conditions (Guerra & Scremin-Dias, 2018).

The presence of leaf hairs or trichomes is a common characteristic of plants in dry environments, where their high density represents an adaptation to limited water availability and high temperatures (Guerra & Scremin-Dias, 2018). The results of our study indicated that out of the six species, only two species possessed trichomes, namely *N. cadamba* and *R. balangeran* (Fig. 6a and 7b). *N. cadamba* had trichomes of the unicellular type (Fig. 6a), while *R. balangeran* had the stellate (star-like) trichomes type (Fig. 6b). Compared to *N. cadamba*, the trichomes on *R. balangeran* were more dense, covering almost the entire abaxial surface (Fig. 6b), while the trichomes on *N cadamba* were more abundant in the leaf veins (Fig. 6a). Study of Araújo et al. (2024) showed that compared to *Croton adenocalyx* with sparse stellate trichomes, *Croton blanchetianus* with dense stellate trichomes exhibits increased leaf gas exchange, improved water status, enhanced capacity to reflect light, and maintaining lower temperatures during drought. This mechanism enables *C. blanchetianus* leaves to absorb water more efficiently than those of *C. adenocalyx*.

Thus, based on the anatomical characteristics, *R. balangeran* demonstrates the highest adaptive capacity to the conditions in the ex-mining area with high sun exposure intensity and dry conditions, followed by *N. cadamba* and *D. costulata*. This is evidenced by the thickening of leaves, cuticle, epidermis (and its derivatives), mesophyll thickening (size and number of palisade layers), the presence of BSE, trichomes, and sclerenchyma. Meanwhile, *A. angustiloba*, *L. artocarpifolia*, and *S. myrtifolium* tend to be more adaptive in waterlogged areas, with thinner leaf characteristics and sparse mesophyll tissue, and the palisade consists of only one layer. Therefore, the emergence of adaptive or tolerant characteristics in certain environmental conditions can serve as a basis for assessing the most adaptive species by determining the most abundant adaptive anatomical characteristics possessed by each species.

Our study in post-mining reclamation area provides a useful insight that plant traits, specifically leaf traits, can be used as a basis for selecting species for restoration, as well as for monitoring and evaluating the success of the restoration efforts. The flow of activities, methodology, and recommendations based on the study conducted can be illustrated in the following diagram (Fig. 7). The leaf trait approach used is not only cost-effective and easily applicable but also accurate in efforts to restore post-mining ecosystems toward a system similar to the reference area in both structure and function that it produces.

5 Challenges and Consideration

In this article, we showcase the use of eco-physiological and anatomical traits of plants for ecological restoration and reclamation of degraded lands in tropical regions. We provide several examples of plants native to Indonesia which can be used to inform land restoration and reclamation in the country. Considering that Indonesia have a great biological diversity in terms of ecoregions, ecosystem types and plant species we acknowledge that the plants presented in this article only resemble a very small portion of large number of Indonesian flora, let alone tropical forests globally which is estimated to contain huge number of tree species. Therefore, future efforts are directed to extensively reveal the eco-physiological and anatomical traits of various plant species especially trees for ecological restoration. While such assessment will add useful knowledge in restoration ecology, there are several challenges and considerations which might exist as elaborated here.

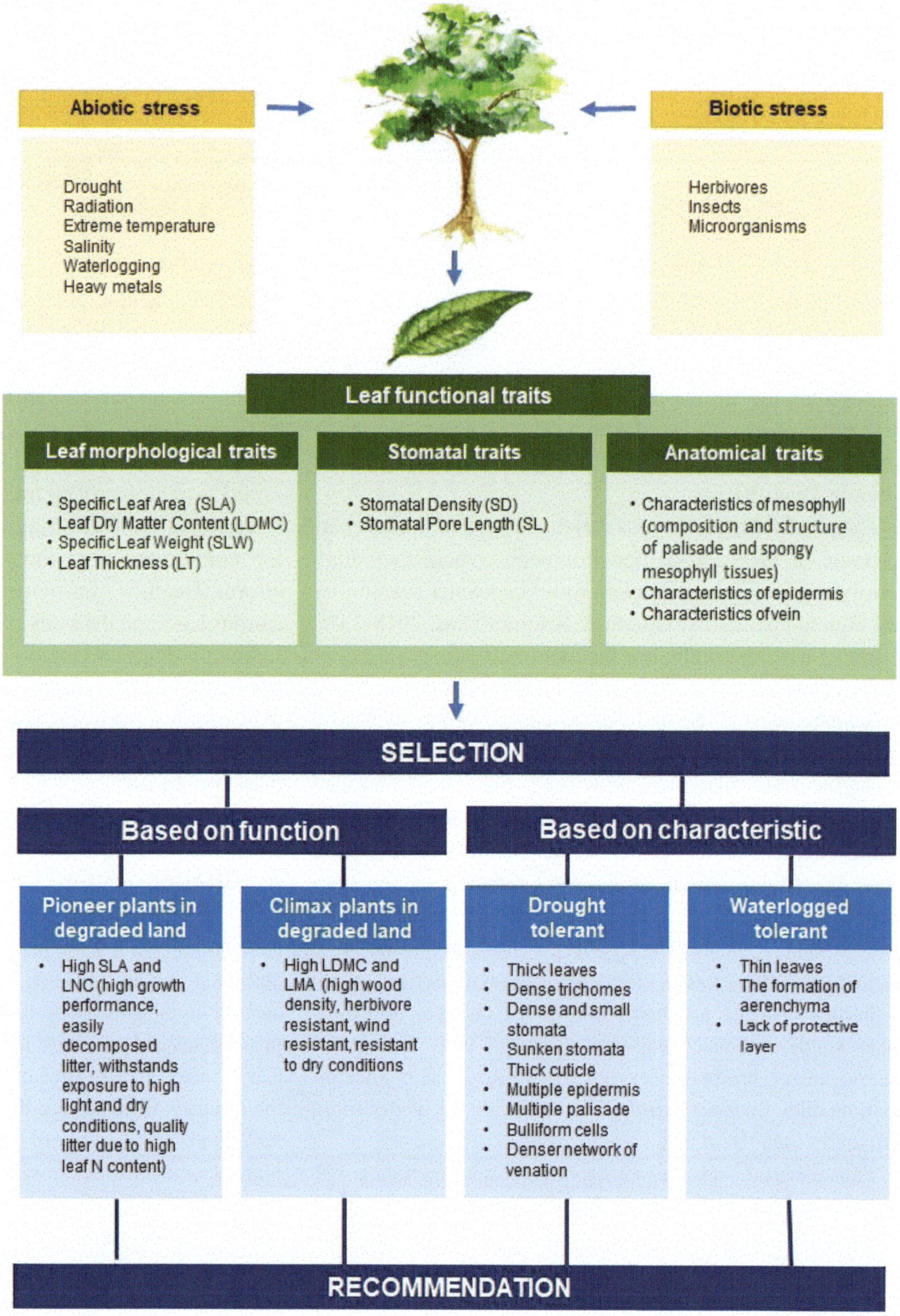

Fig. 7 Analytical framework of using plant ecophysiological and anatomical traits for restoration and reclamation of degraded lands

First, integrating these advanced approaches requires expertise and collaboration between ecologists, plant taxonomists, plant physiologists and anatomists, and restoration practitioners. The approaches also combine works in the field and laboratory analysis with some extent of desktop studies. To be effective, preliminary studies involving reviews of secondary data might be needed to screen a great number of species to produce a manageable list of plants with information on biogeography, potential uses, conservation status and other considerations relevant to land reclamation and restoration. Based on such a list, subsequent works in the field are necessary to investigate the ecological aspects of the plant studied including the micro-climate and soil variables where the plant is occurring. The plant is measured in the field with several specimens, such as leaves and wood samples, are brought for laboratory analyses for investigation on eco-physiological and anatomical traits as explained in the Sect. 2. Further analysis using statistical methods to look at the relationships and effects of ecological variables on eco-physiological and anatomical traits might be useful to draw conclusion whether the plant studied has the potential for restoration and reclamation of degraded lands. Finally, a list of plants resulting from previous stages necessitates field trials before management adoption by practitioners by looking at several aspects including technical applicability and financial feasibility.

Second challenge is related to funding for conducting the fieldwork studies and laboratory analyses. The extensive works needed to reveal the eco-physiological and anatomical traits of numerous plants requires a considerable amount of research funds. Such financial resources are not necessarily available for developing countries such as Indonesia. The lack of research funding is particularly eminent in the field of explorative studies as in the case of investigating plant traits. To resolve this problem, there is a need of partnerships beyond researchers and research funding agencies, which can be extended to involve policy makers and private sectors responsible to land degradation and restoration (Hapsari et al., 2023).

The third challenge is cost and logistical issues related to technical and financial capacities for the widespread implementation of the advanced approaches. Eventually, the end users of these approaches are land restoration and reclamation practitioners, from local communities to government agencies and companies which have a broad range of financial and technical capacity. The final product of the advanced approaches is a list of plant species (probably in a great number) recommended for restoration and reclamation of degraded lands in Indonesia. To some stakeholders including private companies, planting a lot of tree species might be doable due to its financial and technical capacities (Lestari et al., 2019), while other stakeholders such as local communities, might not have such adequate resources.

The last stage of these approaches involves field trials by conducting a long-term monitoring and evaluation to assess the effectiveness of the approaches before the adoption by practitioners. Conducting such ecological studies requires extensive resources ranging from money, labors, lands and plant materials, and a long period of monitoring often up to several decades. Despite the great amount of resources invested and the long process needed, restoration and reclamation of degraded lands have uncertain results which are influenced by various factors, not only the correct selection of species planted (Trimanto et al., 2021). Such challenges inevitably hinder the immediate adoption of the advanced approaches since the practitioners need to see the empirical evidence of restoration and reclamation success.

The application of the advanced approach in land restoration and reclamation would support the implementation and achievement of various pledges, policies and regulations, either at global or national levels. At the global level, the Kunming-Montreal Global Biodiversity Framework states in Target 2 that "by 2030 at least 30 percent of areas of degraded terrestrial, inland water, and marine and coastal ecosystems are under effective restoration, in order to enhance biodiversity and ecosystem functions and services, ecological integrity and connectivity". This target implies the importance of using various species to enhance biodiversity and ecosystem services, thus the understanding of their biological characteristics, including the eco-physiological and anatomical traits, are pre-requisite for the successful restoration to achieve the Target 2.

At national level, there is currently a paradigm shift in policies and regulations in Indonesia where forest rehabilitation is not only focused on increasing vegetation cover but also to enhance biodiversity and ecological resilience. This is particularly eminent for restoration in unique ecosystems, such as peatland, and severely degraded areas, like post-mined land. Restoring peatland is very challenging due to the unique environmental characteristics (e.g., water logged, high organic carbon matter, low fertility), thus it requires specific plant species for restoration (Ismail et al., 2023). To resemble the scale of the problem, Indonesia even creates a specific government body to restore degraded peatland called Peat and Mangrove Restoration Agency. In the context of mining, the Indonesian government puts an obligation on mining operators to plant at least 40% of native species in post-mining reclamation. This is difficult to implement since soil and climate conditions in post-mined land are in extreme conditions which might result in arrested succession if no soil management and active planting is conducted (Albasri et al., 2023). To solve such problems, better understanding of plant traits might be useful to investigate which species can survive and grow in degraded peatlands and mined areas.

References

Ahrens, C. W., Andrew, M. E., Mazanec, R. A., Ruthrof, K. X., Challis, A., Hardy, G., Byrne, M., Tissue, D. T., & Rymer, P. D. (2020). Plant functional traits differ in adaptability and are predicted to be differentially affected by climate change. *Ecology and Evolution, 10*(1), 232–248. https://doi.org/10.1002/ece3.5890

Albasri, A., Tuheteru, F. D., Husna, H., Budiharta, S., Tuheteru, E. J., Mulyono, S., Yunanto, T., & Karlin, V. (2023). Composition, structure and diversity of vegetation on asphalt post-mining sites in Buton Island, Indonesia. *Floresta Ambient, 30*(3), e20230022. https://doi.org/10.1590/2179-8087-FLORAM-2023-0022

Alscher, R. G., Erturk, N., & Heath, L. S. (2002). Role of superoxide dismutases (SODs) in controlling oxidative stress in plants. *Journal of Experimental Botany, 53*(372), 1331–1341. https://doi.org/10.1093/jexbot/53.372.1331

Araújo, K. C., Souza, B. C., Carvalho, E. C. D., Freire, R. S., Teixeira, A. S., Muniz, C. R., Martins F. R., Oliveira, R. S., Eller, C. B. & Soares, A. A. (2024). The multiple roles of trichomes in two Croton species. *Plant, Cell & Environment, 47*(5), 1685–1700. https://doi.org/10.1111/pce.14829

Bahadur, A., Batool, A., Nasir, F., Jiang, S., Minsen, Q., Zhang, Q., Pan, J., Liu, Y., & Feng, H. (2019). Mechanistic insights into arbuscular mycorrhizal fungi-mediated drought stress tolerance in plants. *International Journal of Molecular Sciences, 20*(17), 4199. https://doi.org/10.3390/ijms20174199

Bucher, S. F., Auerswald, K., Grun-Wenzel, C., Higgins, S. I., Jorge, J. G., & Romermann, C. (2017). Stomatal traits relate to habitat preferences of herbaceous species in a temperate climate. *Flora, 229*, 107–115. https://doi.org/10.1016/j.flora.2017.02.011

Carlucci, M. B., Brancalion, P. H., Rodrigues, R. R., Loyola, R., & Cianciaruso, M. V. (2020). Functional traits and ecosystem services in ecological restoration. *Restoration Ecology, 28*(6), 1372–1383. https://doi.org/10.1111/rec.13279

Chaves, M. M., Costa, J. M., Zarrouk, O., Pinheiro, C., Lopes, C. M., & Pereira, J. S. (2016). Controlling stomatal aperture in semi-arid regions-The dilemma of saving water or being cool? *Plant Science, 251*, 54–64. https://doi.org/10.1016/j.plantsci.2016.06.015

Chirino, E., Vilagrosa, A., & Vallejo, V. R. (2011). Using hydrogel and clay to improve the water status of seedlings for dryland restoration. *Plant and Soil, 344*, 99–110. https://doi.org/10.1007/s11104-011-0730-1

Choi, Y. D. (2004). Theories for ecological restoration in changing environment: Toward 'futuristic' restoration. *Ecological Research, 19*, 75–81. https://doi.org/10.1111/j.1440-1703.2003.00594.x

Clement, C., Schneider, H. M., Dresboll, D. B., Lynch, J. P., & Kristensen, K. T. (2022). Root and xylem anatomy varies with root length, root order, soil depth and environment in intermediate wheatgrass (Kernza) and alfalfa. *Annals of Botany, 130*(3), 367–382. https://doi.org/10.1093/aob/mcac058

Clewell, A. F., & Aronson, J. (2013). Ecological restoration: Principles, values, and structure of an emerging profession. *Island Press*. https://doi.org/10.5822/978-1-59726-323-8

Coneva, V., Frank, M. H., de Luis Balaguer, M. A., Li, M., Sozzani, R., & Chitwood, D. H. (2017). Genetic architecture and molecular networks underlying leaf thickness in desert-adapted tomato *Solanum pennellii*. *Plant Physiology, 175*(1), 376–391. https://doi.org/10.1104/pp.17.00790

Dominguez, E., Cuartero, J., & Heredia, A. (2011). An overview on plant cuticle biomechanics. *Plant Science, 181*(2), 77–84. https://doi.org/10.1016/j.plantsci.2011.04.016

Donohue, K., Rubio de Casas, R., Burghardt, L., Kovach, K., & Willis, C. G. (2010). Germination, postgermination adaptation, and species ecological ranges. *Annual Review of Ecology, Evolution, and Systematics, 41*, 293–319. https://doi.org/10.1146/annurev-ecolsys-102209-144715

Evert, R. F. (2006). *Esau's plant anatomy: Meristems, cells, and tissues of the plant body: Their structure, function, and development* (3rd ed.). Jonh Wiley and Son. https://doi.org/10.1002/0470047380

Feng, L., Raza, M. A., Li, Z., Chen, Y., Khalid, M. H. B., Du, J., Liu, W., Wu, X., Song, C., Yu, L., Zhang, Z., Yuan, S., Yang, W., & Yang, F. (2019). The influence of light intensity and leaf movement on photosynthesis characteristics and carbon balance of soybean. *Frontiers in Plant Science, 9*, 1952. https://doi.org/10.3389/fpls.2018.01952

Fiqa, A. P., Yulistyarini, T., Damaiyani, J., Fauziah, Budiharta, S., Hapsari, L., Trimanto, Rahadiantoro, A., Miako, F., & Bachruddin. (2024). *Studi ekofisiologis tumbuhan berkayu hutan dipterocarpaceae kalimantan di area Konsesi PT trubaindo coal mining*. Badan Riset dan Inovasi Nasional.

Food and Agriculture Organization of the United Nations. (2020). *Global forest resources assessment 2020: Main report*. https://doi.org/10.4060/ca9825en

Funk, J. L., Cleland, E. E., Suding, K. N., & Zavaleta, E. S. (2008). Restoration through reassembly: Plant traits and invasion resistance. *Trends in Ecology & Evolution, 23*(12), 695–703. https://doi.org/10.1016/j.tree.2008.07.013

Galli, U., Schuepp, H., & Brunold, C. (1994). Heavy metal binding by mycorrhizal fungi. *Physiologia Plantarum, 92*(2), 364–368. https://doi.org/10.1111/j.1399-3054.1994.tb05349.x

Gastauer, M., Silva, J. R., Junior, C. F. C., Ramos, S. J., Souza Filho, P. W. M., Neto, A. E. F., & Siqueira, J. O. (2018). Mine land rehabilitation: Modern ecological approaches for more sustainable mining. *Journal of Cleaner Production, 172*, 1409–1422. https://doi.org/10.1016/j.jclepro.2017.10.223

Global Forest Watch. (2024). *Location of tree cover gain in Indonesia*. www.globalforestwatch.org

Goodwin, S. M., & Jenks, M. A. (2005). Plant cuticle function as a barrier to water loss. In M. A. Jenks, & P. M. Hasegawa (Eds.), *Plant abiotic stress* (pp. 14–36). Blackwell Publishing Ltd. https://doi.org/10.1002/9780470988503.ch2

Guerra, A., & Scremin-Dias, E. (2018). Leaf traits, sclerophylly and growth habits in plant species of a semiarid environment. *Brazilian Journal of Botany, 41*(1), 131–144. https://doi.org/10.1007/s40415-017-0416-x

Gunarathne, V., Mayakaduwage, S., & Vithanage, M. (2017). Biochar's influence as a soil amendment for essential. In M. Naeem, A. Ansari, & S. Gill (Eds.), *Essential plant nutrients* (pp. 47–67). Springer. https://doi.org/10.1007/978-3-319-58841-4_3

Guo, C. Y., Ma, L. N., Yuan, S., & Wang, R. Z. (2017). Morphological, physiological and anatomical traits of plant functional types in temperate grasslands along a large-scale aridity gradient in northeastern China. *Scientific Reports, 7*, 40900. https://doi.org/10.1038/srep40900

Guo, W., Cherubini, P., Zhang, J., Li, M-H., & Qi, L. (2023). Leaf s Grigore tal traits rather than anatomical traits regulate gross primary productivity of moso bamboo (*Phyllostachys edulis*) stands. *Frontiers in Plant Science, 14*, 1117564. https://doi.org/10.3389/fpls.2023.1117564

Gutjahr, C., & Paszkowski, U. (2013). Multiple control levels of root system remodeling in arbuscular mycorrhizal symbiosis. *Frontiers in Plant Science, 4*, 204. https://doi.org/10.3389/fpls.2013.00204

Hapsari, L., Yulistyarini, T., Danarto, S. A., Mas' udah, S., Damaiyani, J., Laksono, R. A., Jati, A., & Budiharta, S. (2023). Ecological analysis of community and private partnership in tree planting program to rehabilitate degraded lands: A case study in East Java, Indonesia. *Jurnal Manajemen Hutan Tropika, 29*(1), 22–34. https://doi.org/10.7226/jtfm.29.1.22

Hapsari, L. (2022). *Evaluasi performa ekofisiologi jenis pohon terpilih dan implikasinya untuk strategi restorasi ekosistem pegunungan*

tropis: *Studi kasus di Tahura R. Soerjo, Jawa Timur* (Laporan Magang Riset). Badan Riset dan Inovasi Nasional.
Haworth, M., Marino, G., Materassi, A., Raschi, A., Scutt, C. P., & Centritto, M. (2023). The functional significance of the stomatal size to density relationship: Interaction with atmospheric [CO_2] and role in plant physiological behaviour. *Science of The Total Environment, 863*, 160908. https://doi.org/10.1016/j.scitotenv.2022.160908
He, N., Liu, C., Piao, S., Sack, L., Xu, L., Luo, Y., He, J., Han, X., Zhou, G., Zhou, X., & Lin, Y. (2018). Ecosystem traits linking functional traits to macroecology. *Trends in Ecology & Evolution, 34*(3), 200–210. https://doi.org/10.1016/j.tree.2018.11.004
Ismail, I., Haghighi, A. T., Marttila, H., Karyanto, O., & Kløve, B. (2023). Recent results from an ecohydrological study of forest species in drained tropical peatlands. *Agricultural and Forest Meteorology, 331*, 109338. https://doi.org/10.1016/j.agrformet.2023.109338
Jackson, M. B., & Armstrong, W. (1999). Formation of aerenchyma and the processes of plant ventilation in relation to soil flooding and submergence. *Plant Biology, 1*, 274–287. https://doi.org/10.1111/j.1438-8677.1999.tb00253.x
Jaramillo, R. E., Nord, E. A., Chimungu, J. G., Brown, K. M., & Lynch, J. P. (2013). Root cortical burden influences drought tolerance in maize. *Annals of Botany, 112*(2), 429–437. https://doi.org/10.1093/aob/mct069
Jordan, G. J., Dillon, R. A., & Weston, P. H. (2005). Solar radiation as a factor in the evolution of scleromorphic leaf anatomy in Proteaceae. *American Journal of Botany, 92*(5), 789–796. https://doi.org/10.3732/ajb.92.5.789
Kardiman, R., & Raebild, A. (2017). Relationship between stomatal density, size and speed of opening in Sumatran rainforest species. *Tree Physiology, 38*(5), 696–705. https://doi.org/10.1093/treephys/tpx149
Khaliq, A., Perveen, S., Alamer, K. H., Zia, U. I., Haq, M., Rafique, Z., Alsudays, I. M., Althobaiti, A. T., Saleh, M. A., Hussain, S., & Attia, H. (2022). Arbuscular mycorrhizal fungi symbiosis to enhance plant–soil interaction. *Sustainability, 14*(13), 7840. https://doi.org/10.3390/su14137840
Klich, M. G. (2000). Leaf variations in Elaeagnus angustifolia related to environmental heterogeneity. *Environmental and Experimental Botany, 44*(3), 171–183. https://doi.org/10.1016/S0098-8472(00)00056-3
Lack, A. J., & Evans, D. E. (2001). *Instant notes of plant biology*. BIOS Scientific Publishers Limited. https://doi.org/10.1201/9780429258312
Laughlin, D. C. (2014). Applying trait-based models to achieve functional targets for theory driven ecological restoration. *Ecology Letters, 17*(7), 771–784.
Lestari, D. A., Fiqa, A. P., Fauziah, F., & Budiharta, S. (2019). Growth evaluation of native tree species planted on post coal mining reclamation site in East Kalimantan, Indonesia. *Biodiversitas Journal of Biological Diversity, 20*(1), 134–143. https://doi.org/10.13057/biodiv/d200116
Liu, W., Zheng, L., & Qi, D. (2020). Variation in leaf traits at different altitudes reflects the adaptive strategy of plants to environmental changes. *Ecology and Evolution, 10*(15), 8166–8175. https://doi.org/10.1002/ece3.6519
Lynch, J. P. (2013). Steep, cheap and deep: An ideotype to optimize water and N acquisition by maize root systems. *Annals of Botany, 112*, 347–357. https://doi.org/10.1093/aob/mcs293
Magan, N., Sanchis, V., & Aldred, D. (2004). The role of spoilage fungi in seed deterioration. *Mycology Series, 21*, 311–324. https://doi.org/10.1201/9780203913369.ch28
Ministry of Forestry of Indonesia. (2022). *The State of Indonesia's Forests 2022 towards FoLU Net Sink 2030*. https://phl.menlhk.go.id/publikasi/the-state-of-indonesias-forests-2022-towards-folu-net-sink-2030/
Murchie, E. H., & Lawson, T. (2013). Chlorophyll fluorescence analysis: A guide to good practice and understanding some new applications. *Journal of Experimental Botany, 64*(13), 3983–3998. https://doi.org/10.1093/jxb/ert208
Mutia, A. (2022). *Tren Kehilangan Hutan Primer Indonesia Menurun dalam 5 Tahun Terakhir. UpToDate*. Retrieved April 1, 2024, from https://databoks.katadata.co.id/datapublish/2022/11/16/tren-kehilangan-hutan-primer-indonesia-menurun-dalam-5-tahun-terakhir
Nardini, A. (2022). Hard and tough: the coordination between leaf mechanical resistance and drought tolerance. *Flora, 288*, 152023. https://doi.org/10.1016/j.flora.2022.152023
Ni, X., Sun, L., Cai, Q., Ma, S., Feng, Y., Sun, Y., An, L., & Ji, C. (2022). Variation and determinants of leaf anatomical traits from boreal to tropical forests in eastern China. *Ecological Indicators, 140*, 108992. https://doi.org/10.1016/j.ecolind.2022.108992
Pedrini, S., Balestrazzi, A., Madsen, M. D., Bhalsing, K., Hardegree, S. P., Dixon, K. W., & Kildisheva, O. A. (2020). Seed enhancement: Getting seeds restoration-ready. *Restoration Ecology, 28*, S266–S275. https://doi.org/10.1111/rec.13184
Penfield, S. (2017). Seed dormancy and germination. *Current Biology, 27*(17), R874–R878. https://doi.org/10.1016/j.cub.2017.05.050
Poejiastuti, E. (1994). *Studi komparatif anatomi Daun Beberapa Genotif Kedelai (Glycine max L.) yang Peka dan Toleran terhadap Cekaman Kekeringan*. [Skripsi tidak dipublikasikan]. Institut Pertanian Bogor.
Poorter, H., Niinemets, Ü., Poorter, L., Wright, I. J., & Villar, R. (2009). Causes and consequences of variation in leaf mass per area (LMA): A meta-analysis. *New Phytologist, 182*(3), 565–588. https://doi.org/10.1111/j.1469-8137.2009.02830.x
Potapov, P., Hansen, M. C., Pickens, A., Hernandez-Serna, A., Tyukavina, A., Turubanova, S., Zalles, V., Li, X., Khan, A., Stolle, F., Harris, N., Song, X-P., Baggett, A., Kommareddy, I., & Kommareddy, A. (2022). The global 2000–2020 land cover and land use change dataset derived from the landsat archive: First results. *Frontiers Remote Sensing, 3*, 856903. https://doi.org/10.3389/frsen.2022.856903
Prasad, R., Bhola, D., Akdi, K., Cruz, C., KVSS, S., Tuteja, N., & Varma, A. (2017). Introduction to mycorrhiza: Historical development. In A. Varma, R. Prasad, & N. Tuteja (Eds.), *Mycorrhiza-function, diversity, state of the art*. Springer. https://doi.org/10.1007/978-3-319-53064-2_1
Priori, S., Pellegrini, S., Vignozzi, N., & Costantini, E. A. C. (2021). Soil physical-hydrological degradation in the root-zone of tree crops: Problems and solutions. *Agronomy, 11*(1), 68. https://doi.org/10.3390/agronomy11010068
Puigdefabregas, J. (2005). The role of vegetation patterns in structuring runoff and sediment fluxes in drylands. *Earth Surface Processes Landforms, 30*(2), 133–147. https://doi.org/10.1002/esp.1181
Ramakrishnan, M., Papolu, P. K., Satish, L., Vinod, K. K., Wei, Q., Sharma, A., Emamverdian, A., Zou, L. H., & Zhou, M. (2022). Redox status of the plant cell determines epigenetic modifications under abiotic stress conditions and during developmental processes. *Journal of Advanced Research, 42*, 99–116. https://doi.org/10.1016/j.jare.2022.04.007
Rindyastuti, R., Nurfadilah, S., Rahadiantoro, A., Hapsari, L., & Abywijaya, I. K. (2018). Leaf anatomical characters of four epiphytic orchids of Sempu Island, East Java, Indonesia: The importance in identification and ecological adaptation. *Biodiversitas, 19*(5), 1906–1918. https://doi.org/10.13057/biodiv/d190543
Rollon, R. J. C., Galleros, J. E. V., Galos, G. R., Villasica, L. J. D., & Carcia, C. M. (2017). Growth and nutrient uptake of Paraserianthes falcataria (L.) as affected by carbonized rice hull and arbuscular mycorrhizal fungi grown in an artificially copper contaminated soil. *AAB Bioflux, 9*(2), 57–67. http://www.aab.bioflux.com.ro/home/volume-9-2-2017/
Roth, I. (1984). *Stratification of tropical forests as seen in leaf structure*. W. Junk Publisher. https://doi.org/10.1007/978-94-009-6569-0

Saturday, A. (2018). Restoration of degraded agricultural land: a review. *Journal of Environment and Health Science, 4*(2), 44–51. https://www.ommegaonline.org/article-details/Restoration-of-Degraded-Agricultural-Land-A-Review/1928

Schenk, H. J., & Jackson, R. B. (2002). Rooting depths, lateral root spreads and below-ground/above-ground allometries of plants in water-limited ecosystems. *Journal of Ecology, 90*(3), 480–494. https://doi.org/10.1046/j.1365-2745.2002.00682.x

Sharma, I., & Ahmad, P. (2014). Chapter 4–Catalase: A versatile antioxidant in plants. In P. Ahmad (Ed.), *Oxidative damage to plants* (pp. 131–148). Academic Press. https://doi.org/10.1016/B978-0-12-799963-0.00004-6

Siefert, A., Violle, C., Chalmandrier, L., Albert, C. H., Taudiere, A., Fajardo, A., Aarssen, L. W., Baraloto, C., Carlucci, M. B., Cianciaruso, M. V., & de L. Dantas, V. (2015). A global meta-analysis of the relative extent of intraspecific trait variation in plant communities. *Ecology Letters, 18*(12), 1406–1419. https://doi.org/10.1111/ele.12508

Solberg, S. Ø., Yndgaard, F., Andreasen, C., & Von Bothmer, R. (2020). Long-term storage and longevity of orthodox seeds: A systematic review. *Frontiers in Plant Science, 11*, 548383. https://doi.org/10.3389/fpls.2020.01007

Suding, K. N. (2011). Toward an era of restoration in ecology: Successes, failures, and opportunities ahead. *Annual Review of Ecology, Evolution, and Systematics, 42*, 465–487. https://doi.org/10.1146/annurev-ecolsys-102710-145115

Taiz, L., & Zeiger, E. (2002). *Plant physiology* (3rd ed.), Sinauer Associates, Inc.

Trimanto, T., Hapsari, L., & Budiharta, S. (2021). Integrating indicators of natural regeneration, enrichment planting, above-ground carbon stock, micro-climate and soil to assess vegetation succession in post-mining reclamation in tropical forest. *Turkish Journal of Botany, 45*(5), 457–467. https://doi.org/10.3906/bot-2103-35

Walters, C. (2015). Orthodoxy, recalcitrance and in-between: Describing variation in seed storage characteristics using threshold responses to water loss. *Planta, 242*, 397–406. https://doi.org/10.1007/s00425-015-2312-6

Wang, G., Ren, Y., Bai, X., Su, Y., & Han, J. (2022). Contributions of beneficial microorganisms in soil remediation and quality improvement of medicinal plants. *Plants, 11*(23), 3200. https://doi.org/10.3390/plants11233200

Widyati, E. (2006). *Ex-coal mining land bioremediation using sludge of paper mills to enhance land revegetation* [Doctoral dissertation, IPB University]. IPB Repository. http://repository.ipb.ac.id/handle/123456789/46957

Wu, J. (2013). Landscape sustainability science: Ecosystem services and human well-being in changing landscapes. *Landscape Ecology, 28*, 999–1023. https://doi.org/10.1007/s10980-013-9894-9

Xu, J., Zhang, Z., Zhao, X., & Cheng, S. (2023). Downward trend of NO2 in the urban areas of Beijing-Tianjin-Hebei region from 2014 to 2020: Comparison of satellite retrievals, ground observations, and emission inventories. *Atmospheric Environment, 295*, 119531. https://doi.org/10.1016/j.atmosenv.2022.119531

Yulistyarini, T., Fiqa, A. P., Damaiyani, J., Fauziah, Budiharta, S., Hapsari, L., Trimanto, Rahadiantoro, A., Miako, F., & Bachruddin. (2024). *Penelitian bidang konservasi diversitas tumbuhan lokal di kawasan konsesi pertambangan kabupaten kutai barat, Kalimantan Timur*. Badan Riset dan Inovasi Nasional.

Zhang, Y., Wang, L. F., Li, T. T., & Liu, W. C. (2021). Mutual promotion of LAP2 and CAT2 synergistically regulates plant salt and osmotic stress tolerance. *Frontiers in Plant Science, 12*, 672672. https://doi.org/10.3389/fpls.2021.672672

Zhou, M., Guo, Y., Sheng, J., Yuan, Y., Zhang, W. H., & Bai, W. (2022). Using anatomical traits to understand root functions across root orders of herbaceous species in a temperate steppe. *New Phytologist, 234*(2), 422–434. https://doi.org/10.1111/nph.17978

Zhu, J. M., Brown, K. M., & Lynch, J. P. (2010). Root cortical aerenchyma improves the drought tolerance of maize (*Zea mays* L.). *Plant, Cell and Environment, 33*(5), 740–749. https://doi.org/10.1111/j.1365-3040.2009.02099.x

Zirbel, C. R., Bassett, T., Grman, E., & Brudvig, L. A. (2017). Plant functional traits and environmental conditions shape community assembly and ecosystem functioning during restoration. *Journal of Applied Ecology, 54*(4), 1070–1079. https://doi.org/10.1111/1365-2664.12885

Unveiling the Dynamic Role of Microbes in Soil Management for Sustainable Agriculture

Nagma Khan, Baby Tabassum, Mohammad Hashim, and Asma Hasan

Abstract

The implementation of sustainable agricultural practices is highly dependent on the intricate interrelationships that exist among microorganisms. This chapter explores the crucial role of microorganisms in maintaining soil health and avoiding soil-borne diseases. Microbial communities are essential for providing nutrients that enhance plant development and soil ecosystems. The continuous expansion of the human population poses a threat to sustainable agriculture by disrupting the complex equilibrium of soil microbiota, leading to land degradation and a reduction in microbial populations. Some microbial groups participate in nutrient cycling, while others use their biochemical abilities to enhance plant resistance to diseases and pests. Utilizing these microbial partners improves agricultural output while also supporting environmental sustainability and human health. This synthesis explores several microbial organisms, such as plant growth-promoting rhizobacteria (PGPR) and bioremediation specialists, highlighting their varied roles in improving soil health. Innovative approaches that combine advanced biotechnologies with traditional agricultural methods show potential for improving the effectiveness of microbes in soil management. Innovative methods, such as precision micro biome engineering and customized microbial inoculants, are becoming guiding lights for sustainable soil management. We use the diverse soil microbial community to create strong agricultural ecosystems and promote an environmentally friendly and better planet for future generations.

Keywords

Sustainable agriculture · Soil health · Microbial diversity · Plant–microbe interactions · Bioremediation · PGPR

1 Introduction

Soil is a limited and valuable resource that sustains many land-based organisms due to its essential activities. Soil quality is the ability of soil to concern as a crucial natural system within certain bounds, supporting biological productivity, enhancing air and water quality, and preserving the health of plants, animals, and humans. Soil health in sustainable agriculture refers to its role in enhancing crop productivity and protection through functions like N_2 fixation and phosphorus (P) solubilization, maintaining biogeochemical cycles, preserving soil composition, detoxifying impurities, and suppressing plant pathogens. Without these processes, the soil is considered an inanimate object of inorganic materials and chemicals. Soil restoration by chemical and biological processes, as well as the weathering of the underlying rock, is a time-consuming geological process (Singh et al., 2016). Soil health is determined by physical variables like permeability, humidity, and appearance; chemical variables including decaying organisms, substances, carbon, and nitrogen; and biological variables such as the variety of microbes, the respiration of soil, and the biomass of microbial organisms. Natural factors such as precipitation, soil erosion, wind damage, floods, and landslides, as well as human activities like mining, deforestation, chemical-based agriculture, urbanization,

N. Khan · M. Hashim · A. Hasan
Department of Zoology, Toxicology Lab, Govt. Raza P.G. College, Rampur, UP, India

B. Tabassum (✉)
Mahatma Jyotiba Phule Rohilkhand University, Bareilly, UP, India
e-mail: dr.btabassum@gmail.com

M. Hashim
Department of Biochemistry, Mohammad Ali Jauhar University, Rampur, UP, India

J. A. Parray (ed.), *Soil and Land Use Change*, Sustainable Landscape Planning and Natural Resources Management,
https://doi.org/10.1007/978-3-031-93075-1_10

acidification, alkaline conditions, salinization, and oil spills can cause soil damage. The current priority is to develop strategies for promptly addressing the harm caused by them. Soil health is primarily determined by agricultural output and the ability of the soil endure stressful conditions, suggesting that these goals can only be attained through actions aimed at restoring soil health. Human health and soil health are interconnected. Soil health helps reduce human exposure to disease causing microorganisms, supplies essential components and quality medication, and boosts immunity. The dynamic soil microorganisms help mitigate the severe impacts of climate variation by sequestering more CO_2, therefore alleviating the effects of global warming (Girija Veni et al., 2020). The economic health and crop yield of a country rely on soil assets, and their loss can result in the creation of unproductive barren regions (Bunea and Pacurar, 2020). Both eliminating harmful pollutants and providing nutrients to enhance its overall condition may improve soil health. Utilizing microorganisms that enhance soil fertility by supplying nutrients and neutralizing hazardous substances can restore soil health. Microorganism plays a vital role in agriculture by aiding in insect control, improving nutrient absorption, and serving as bio-fertilizers. They are dynamic, self-sustaining, and self-replicating, removing the necessity for repeated inoculation. These inoculants' efficacy and durability are dependent on environmental factors, plant type, and microbial makeup. Plant micro-biomes contain beneficial microorganisms that enhance plant development, nutrient uptake, nitrogen fixation, and create siderophores to improve crop nutrition. Plant microbiomes play an essential role in agriculture by providing beneficial microorganisms that can promote plant development and boost nutrient intake by solubilizing P, K, and Zn, fixing N_2, and forming siderophores (Prasanna et al., 2016). Beneficial microorganisms can enhance agricultural productivity, eliminate pollutants, suppress harmful microorganisms, and generate nitrogen or new compounds. Plant microbial communities can stimulate growth via variety of mechanisms, including nitrogen fixation by bacteria, the synthesis of growth-promoting substances like gibberellin, inflammatory mediators, and antibiotics; biological control of pathogenic plants by producing antibiotics, antifungals, or antimicrobial substances; the synthesis of Fe-chelating substances; nutrient competition; the emergence of obtained host resistance; and improving the absorption and metabolism of minerals. The study of the intricate relationships between roots of plant and microbial colonies, such as rhizobacteria and fungi, has prompted investigations into the involvement of soil microbes in bio fortification. This process involves enriching crops or foods with vital micronutrients and health-promoting substances to improve their nutritional content. This method is becoming more common to enhance the presence of selenium, Fe, and Zn in important food crops.

Microorganisms with beneficial traits can benefit human health and organic farming. The friendly microbes may serve as supplements in healthy meals for human health. The gut is the primary location for micro biota in the human body, and their makeup is affected by food and dietary choices. The word "micro biota" refers to the collection of living microorganisms in a certain habitat, including soil microbes that control immune responses and hormones and are essential in assessing the nutritional value of food. Yet, the micronutrient levels in many crops are insufficient for human nutrition, leading to cereals and cereal derivatives being the primary meals in many human diets. These foods have the potential to cause serious health problems as well as micronutrient deficiency (Suman et al., 2022). Probiotic bacteria, including the lactic acid bacteria *Lactobacillus*, *Methano brevibacter*, *Methanosphaera*, and *Saccharomyces*, are more often included in functional food items as nutritional supplements. There is a growing interest in creating novel functional foods and incorporating them into a nutritious diet. These products, particularly probiotics, have a positive impact on the micro biota in the gut of the host after ingestion and may be effective in preventing certain illnesses. Probiotics are living microorganisms that, when consumed in sufficient quantities, provide a health advantage to the host (Suman et al., 2022). This review discusses the significance of microorganisms and biotechnology interventions in agricultural sustainability, focusing on their effects on soil quality, ecosystem wellness, and human well-being.

2 The Role of Diverse Microbes in the Plant–Soil System

The collaboration within plants and microbial populations is greatly influenced by various agricultural approaches within the current global framework of agricultural advancement and biological factors. It is crucial to take into account the impact of sustainable stress factors since they have a significant effect on the efficient management of the interactions between crops and the microbiome (Zolla et al., 2013). Microorganisms are the dominant species in soil, exhibiting a substantial level of variety as measured by the diversity index. Nevertheless, the microbial groupings exhibit changes in their abundance relative to the total biomass. Fungi provide superior resilience and adaptation to changing soil conditions, such as those seen in rotating or no-till agricultural soils, in comparison to bacteria and actinomycetes. It is important to mention that both bacteria and actinomycetes are capable of tolerating such situations as well. Soil bacteria are essential in the biological processes of micronutrient cycles, including Carbon, Nitrogen, Sulfur, Phosphorous, and several nutrient cycles. Furthermore, they enhance plant defense against diseases by eradicating soil bacteria through the use of diverse compounds that possess antibiotic resistance. Under regulated conditions, a variety of microbes that play an essential role in ecosystem functioning may be discovered inhabiting the soil. The microorganisms include bacteria, archaea, and fungi. Most bacteria, fungi, and other microorganisms found in soil have a positive impact on plants and the overall condition of the soil. According to Zolla et al. (2013), the number of these microbes in each milligram of soil is expected to be between 10,000 and 50,000. Being close to plant roots enhances the speed at which minerals are absorbed, decomposes organic detritus, enhances the nutrient uptake, promotes plant growth, and suppresses the growth of plant diseases (Nihorimbere et al., 2011).

Microorganisms have a clear and very beneficial function in enhancing soil fertility. Commonly, individuals believe that microorganisms are only responsible for causing diseases. However, it is important to acknowledge that bacteria also engage in several beneficial activities within the biosphere. A great diversity of bacteria, fungus, algae, viruses, protozoa and other microorganisms found in fertile soils. Microorganisms in the soil's organic material are globally utilized to boost plant efficiency. The microorganisms mentioned, such as *Azotobacter, Azospirillum, Bacillus, Klebsiella sp.,* and *Rhizobium sp.,* are beneficial for mitigating issues associated with agricultural chemicals and insecticides. They are widely used in organic and biological farming. The existence of biomass from microorganisms plays a vital role in the storage, conversion, and recycling of substances since it is an active part of biological matter. The soil microbes assist to the movement of nutrients by decomposing decaying matter and supporting ecological processes, including the recycling of nutrients and resistance to environmental and biological stress. A sustainable farming system utilizes many types of microorganisms in the micro biome to acquire nutrients via interactions in the rhizosphere, endophytic, or phyllo sphere, either in symbiotic or free-living ways. These microorganisms also provide plant defense by acting as biopesticides or bio fungicides. The helpful bacteria aid in the breakdown of harmful waste and other contaminants, ultimately contributing to the enhancement of soil fertility. Primarily, they inhabit the rhizosphere, which refers to the soil area immediately around plant roots. Researchers advocate using rhizosphere bacteria over artificial fertilizers to sustain plant development and enhance soil fertility. In this environment, their main function is to break down organic materials, such as plant and animal remains, human waste, and deceased microorganisms, into humus. Humus consists of thoroughly degraded organic materials. Organic matter consists of cellulose, hemicellulose, sucrose, lignin, wax, nucleic acid, protein, and amino acid units (Andreote et al., 2014). Enhancing plant performance is now a multifaceted process that entails contact with certain microorganisms or groups of bacteria. Emerging methodologies include the manipulation of symbiotic associations to viable non legumes and other essential plant material to perform N_2 fixation (Ryu et al., 2020), transforming them into plants that enhance soil fertility (Fig. 1). This would greatly enhance the worldwide food supply and contribute to achieving sustainability objectives.

3 The Agricultural Sector and Plant Microbiota

Soil serves as a natural substrate for the growth and development of plants. The soil is a very varied and intricate environment that harbors a wide range of microorganisms, such as bacteria, actinomycetes, fungi, algae, and more. The plant microbiota plays a crucial role in determining the health and production of plants (Hunter, 2016) and has garnered significant interest in recent times (Trivedi et al., 2017). The rhizosphere bacteria can establish either positive, neutral, or harmful relationships with host plants. The particular interactions between microbes and model plants are well understood. A portion of the results of photosynthesis, ranging from 5 to 20%, is discharged into the rhizosphere via the roots. These products include rhizodeposition exudates, mucilage, and sloughed cells. Root exudates consist of a diverse range of substances, including organic acids, sugars, amino acids, fatty acids, vitamins, growth factors, hormones, and antibacterial agents (Kumar et al., 2019). The composition of the rhizosphere microbiome is influenced by the exudates produced by plants, which might differ depending on the plant's different species, crops, maturity, and phase of development (Compant et al., 2019). The microbiome is very important to plants because it helps them get nitrogen and phosphorus. Nitrogen-fixing bacteria, like rhizobia and *arbuscular mycorrhizal fungi*, and phosphate-solubilizing bacteria work together to make this happen. These bacteria fix nitrogen and break down phosphorus ions, which are then carried to the plant by fungus. Some types of bacteria that can break down phosphate are *Azotobacter* species, *Microbacterium*, *Erwinia*, *Beijerinckia*, *Enterobacter*, *Flavobacterium*, *Pseudomonas*, and *Rhizobium*. *Rhizobia* is part of the legume nodule microbiome, which also has other endophytic microbes that help plants proliferate. Cyanobacteria, *Azotobacter*, *Azospirillum*, and *Gluconacetobacter* are all examples of endophytes that fix nitrogen. Fixators can also live on their own and settle in

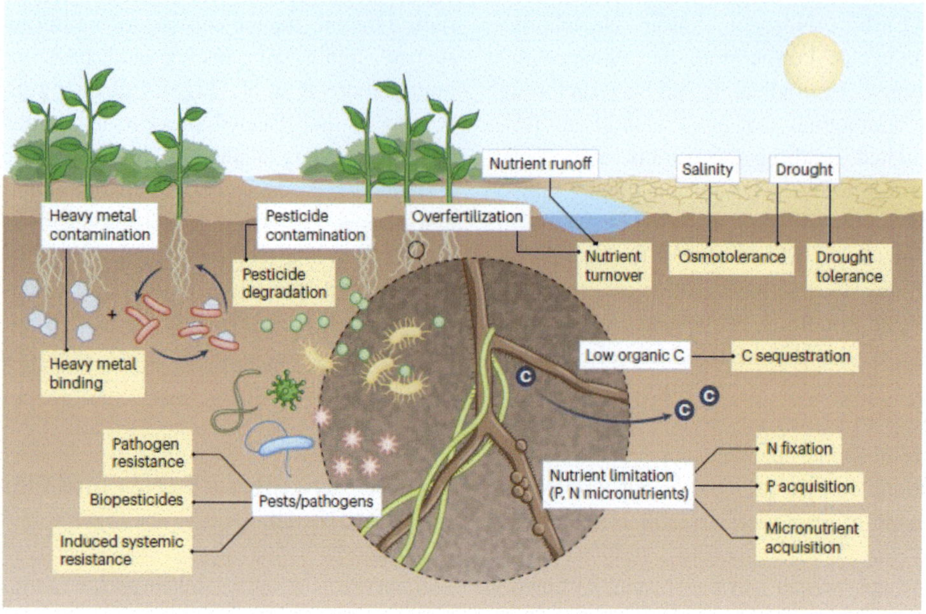

Fig. 1 Soil microorganisms are essential for ecosystem services and can assist in reducing the negative effects of human activities on soil ecosystems. They perform essential functions and have the ability to break down toxic substances such as heavy metals and pesticides. Mycorrhizal fungi, nitrogen-fixing bacteria, and other microbes connected to the roots characterize the rhizosphere, a soil region. The rhizosphere has the ability to break down pesticides, whereas possible harmful organisms include viruses, bacteria, and nematodes. The essay emphasizes the significance of rhizosphere microorganisms in tackling these problems, (*Source* Jansson et al., 2023)

plants that aren't legumes, like *Beijerinckia, Klebsiella*, and *Bacillus* (Kumar et al., 2015).

An intact and well-functioning microbiome is crucial for the ability of plants to withstand and adapt to difficult environmental conditions, including excessive salt levels, a lack of water, and exposure to harmful compounds. High levels of soil salt may hinder the growth and yield of plants. However, the microbiome may counteract this problem by producing phytohormones, which improve the ability of plants to withstand these difficult circumstances. Phytostimulation, or bio stimulation, is the process by which the microbiome generates phytohormones to enhance plant growth. Bacteria produce hormones such as indole-3-acetic acid (IAA), auxin, gibberellins, cytokinin, and salicylic acid. Specific bacteria secrete enzymes that reduce the concentration of ethylene in plants. Studies have shown that phytostimulation can augment plant growth, reduce the need for pesticides, and diminish greenhouse gas emissions. This facilitates the implementation of sustainable agricultural practices and contributes to the survival of the expanding global population (Velazquez et al., 2017). The microbiome's potential in agriculture may be harnessed by using it as an inoculant or manipulating it to choose more effective microbial groups for plant growth. In addition, it is crucial to decrease the use of pesticides and fertilizers by leveraging knowledge about the potential of the plant microbiome. This is essential for promoting sustainable agriculture practices (Table 1).

3.1 The Role of Archaea

The Archaea domain, which contains a wide range of microorganisms, can improve agricultural output by facilitating the recycling of nutrients. These global populations, which are present in diverse environments, have a vital function in the process of nitrogen, sulfur, and carbon recycling, which is necessary for the growth of plants. In addition, they facilitate the absorption of essential nutrients such as nitrogen, phosphorus, sulfur, and potassium, which are crucial for both plant development and soil productivity (Mehta and Baross, 2006). Archaea domains, which rank as the third-largest domain in soil, can enhance plant development, aid in the biogeochemical cycle, and protect against diseases. Their varied habitats and substantial nitrogen cycling enhance agricultural output.

Plant growth-promoting archaea (PGPA) have been noticed to exhibit the capacity to enhance plant growth and boost immunity. This is accomplished through diverse processes, for instance the synthesis of siderophores, which

Table 1 The role of microorganisms with distinct mechanisms and significant actions

Microorganism	Significance of microbe	Mechanism of action	References
Pseudomonas *Azotobacter*	Microorganisms contribute to soil formation by creating organic material and facilitating nutrient cycling	Nitrogen fixation	Aseri et al. (2008)
Bacillus licheniformis K11	Bacillus licheniformis K11 has the ability to successfully enhance resistance to both biotic and abiotic stressors in the cultivation of pepper plants	Formation of auxin and ACC deaminase reduces stress due to drought	Johns et al. (2016)
Mycorrhizal Fungi	Microbes performs immobilization and mineralization	There are two routes in which mycorrhizal plants absorb nutrients from soil: one is known as plant pathway, which uses the outermost layer of the root and hairs for direct uptake, and the other is the mycorrhizal system, which uses fungal ERM transport	De Corato (2020)
Mesorhizobium sp	In soils polluted with metals, encourage the growth of legumes and supply plants with nitrogen	Optimize nutrient absorption and maximize crop productivity	Tardif et al. (2016)
Erwinia species and P. chlororaphis sp.	In order to prevent plant diseases caused by fungi, the root-dwelling bacterium Pseudomonas chlororaphis produces phenazine compounds	The process involves the synthesis of organic substances and the enhancement of the conversion of insoluble phosphate into soluble phosphate within the soil	Bar On et al. (2018)
Lactobacilli, Streptococci	Microbes help to decompose sewage	They enhance the barrier function of the intestinal mucosa, hence inhibiting the attachment of harmful bacteria to host cells	Kacem et al, (2009)
Paenibacillus polymyxa	Biofertilizer, biocontrol agent, PGPR, biofertilizer	In order to improve plant development and resistance to diseases, P. polymyxa is said to produce and release hormones that promote plant growth, such as IAA and soluble inorganic phosphates	Moussa et al, (2013)
Pseudomonas aeruginosa FP6 sp.	Under conditions of iron stress, P. aeruginosa FP6 produces a high siderophore, which chelates iron from the soil and causes ligand exchange with plants. This leads to a negative impact on the proliferation of fungal pathogens	Enhancing siderophore synthesis to improve iron accessibility	Lim and Kim (2013)
Bacillus, Trichoderma	Microbes that help plants thrive by making more nutrients available	Trichoderma modulates pathogenicity via mycoparasitism, antibiotic production, secondary metabolite secretion, and competition for nutrients and space. Bacillus subtilis increases soil P solubilization, nitrogen fixation, and siderophores to grow and prevent infections	Hashem et al. (2019)

(continued)

Table 1 (continued)

Microorganism	Significance of microbe	Mechanism of action	References
Rhizobium, Azospirillum	The soil contains microorganisms that synthesize enzymes and antibacterial substances, which are utilized for biotechnological applications	Azospirillum plants create stress through various strategies, such as systemic tolerance, the formation of hormones and antioxidants, osmotic regulation, and the defense expression of genes	(Kacem et al. 2009)
Streptomyces, Luteimonas, and Enterobacter sp.	Their metabolites may be effective in regulating fungal and bacterial plant pathogens	Inoculants modify the bacterial cell wall to stimulate plant development	Timmusk et al. (2014)

aid in iron acquisition; the solubilization of phosphorus; the fixation of nitrogen by methanogens; and the synthesis of auxins, which are plant hormones that regulate growth and development (Dignam et al., 2018). Nevertheless, the investigation of their contribution to disease resistance has been limited. It has been observed that PGPA can trigger systemic resistance (ISR) against plant pathogenic bacteria. This effect is achieved through a signaling pathway that does not rely on salicylic acid, which is a known regulator of plant defense responses. The ability of *Nitroso cosmicus oleophilus* MY3 determinants to induce ISR (induced systemic resistance) against both necrotrophic and biotrophic pathogens has been observed through their colonization of Arabidopsis thaliana roots. Archaea, a group of microbes, have been found to possess the ability to withstand and endure in tough environmental conditions. They achieve this by applying various mechanisms, such as the formation of Osmo protectants and antioxidant enzymes. Osmo protectants are molecules that help maintain cellular osmotic balance and protect against osmotic stress. These molecules function by controlling the transport water throughout the cell surface. and preventing cellular dehydration. Additionally, archaea also produce antioxidant enzymes that that are vital in preventing the ROS-induced damage.

3.2 The Importance of Fungus

Soil has a diverse array of advantageous and detrimental fungus that impact plants at every stage of their life cycle. An estimated more than one million varieties of fungi are thought to occur in natural environments, although only 5–15% of these have been formally identified and described. Soil fungus often dominates soil ecosystems and make up the biggest part of biomass. They are acknowledged as important contributors to the cycling of nutrients in soil, particularly in processes such as N_2 mineralization, immobilization, and transformation. Mycorrhizae, a kind of fungus, may be found in 80% of land-dwelling plants and have beneficial effects on natural ecosystems. There are two clearly different types of mycorrhizae: *ectomycorrhizae* and *endomycorrhizae.* These categories play a crucial role in advancing sustainable agriculture. *Endomycorrhizae,* namely *arbuscular mycorrhiza* (AM), participate in a beneficial relationship with plants and have the ability to consume around 5 billion tons of carbon each year. Arbuscular mycorrhizal (AM) fungus generates spores within the soil that have the capacity to sprout even in adverse circumstances and assist plants in dealing with abiotic stresses. The symbiotic relationship with plants might potentially aid in reducing carbon emissions and promoting advancements in renewable agricultural practices. Both types of endomycorrhiza are essential for advancing sustainable agriculture. Fungal communities play a vital role in agriculture and the soil environment by producing bioactive substances that are beneficial for plant development and function as agents for controlling pests and diseases. AMF are unique creatures known for their extensive distribution in soil and their essential element in enhancing plant nutrition and well-being, especially under demanding climatic circumstances including fluctuating climates and nutrient-depleted soils. Arbuscular mycorrhizal (AM) fungi act as a mediator between the soil and plants, enabling the transfer of nutrients via symbiotic interactions. While being present in eighty percent of plants, their ubiquity may have been diminished and their composition changed due to farming practices and soil treatments. Fungi may bolster plant resistance to the effects of environmental factors, such as contaminants, saltwater, and dryness, which can be mitigated by the provision of nutrients, enhancement of water intake, and regulation of stress hormones (Wendimu et al., 2023). Using fungal inoculants may enhance agricultural output, reduce dependence on chemical herbicides and insecticides, and improve soil state by enhancing plant tolerance to stress and toxicity.

3.3 The Role of Bacteria

Bacteria are known to have a substantial effect on the synthesis of polysaccharides, which have the ability to bind sand, silt, and clay particles together. This process leads to the formation of microaggregates, ultimately resulting in the enhancement of soil structure. It has been observed that various bacterial species contribute to the formation of an extensive variety of chemical substances. These chemicals are essential for facilitating the binding of particles, leading to the formation of stable microaggregates. Furthermore, this process significantly enhances the overall structure of the soil. Bacteria are essential in the composting process as they transform biodegradable organic materials and garbage into stable nourishment for their own needs. This procedure facilitates the recycling of agricultural waste, enhances the physicochemical qualities of soil, and introduces vital nutrients. Bacteria are an essential component in soil state maintenance by producing enzymes that break down pesticides, plastics, and other contaminants in the soil. This process is called biodegradation. The microbial community, with bacteria and fungi being the predominant and responsible organisms, carries out the biodegradation process. *Bacillus subtilis* and *Pseudomonas fluorescens* are the primary organisms responsible for breaking down agricultural residues, particularly under conditions of elevated moisture. Fungi often become the predominant organisms at the last stage. This process enhances soil fertility, augments microbial variety, boosts population, and improves soil health. According to Thakur (2012), Bacillus amylolytic us outperforms other bacterial species when it comes to degrading plastic, with a rate of 30% weight loss each month. To break down complicated substrates into simpler products, bacteria use a process called mineralization to destroy pesticides and co-metabolisms. Bacteria play a crucial role in maintaining soil health by effectively immobilizing and retaining nutrients within their cellular structures. This phenomenon serves to mitigate the loss of essential nutrients from both the rhizosphere, which is the soil region surrounding the roots, and the rhizoplane, which is the root surface. Furthermore, it is worth noting that nitrogen fixation bacteria play a crucial role in the process of adding atmospheric nitrogen to the soil via biological nitrogen fixation. This process ultimately leads to an increase in the amount of soil nitrogen that is readily available for plants to uptake. During this particular process, the soil bacteria engage in the decomposition of plant leaves and overall biomass. As a result, organic carbon and various advantageous nutrients are released. The processes are essential for maintaining the condition of the soil (Khan et al., 2024a).

Nitrogen is essential for the generation of food, the growth of plants, and the manufacture of biological enzymes. Nitrogen fixation in rhizobia bacteroid facilitates the nodulation of legumes. At present, biological nitrogen fixation is responsible for using 65% of nitrogen in agriculture, which is crucial for maintaining sustained agricultural processes. Rhizobia, a bacterial group, engage in mutualistic symbiosis with legume plants to perform nitrogen fixation (Table 1) (De-Bruijn, 2015). Diverse microbial taxa possess genes encoding nitrogenase, and legumes' mutualistic association with bacteria that fix nitrogen results in noteworthy metabolic processes that transform elemental nitrogen into ammonia.

3.4 The Impact of PGPR on Soil Quality is an Important Consideration

Microorganisms inhabit the soil naturally and have a vital impact on the functioning of the soil and the production of plants. To achieve effective use of microbial bio inoculants and their influence on the condition of the soil, scientific investigations have emphasized the comprehension of indigenous microbial communities' variety in soil and their spatial distribution. PGPR, also known as plant growth promoters, performs a range of advantageous functions in the soil. These include breaking down crop residues, converting organic matter into minerals, retaining nutrients, enhancing phosphate solubility, facilitating nitrification, fixing nitrogen, promoting plant growth through bio fertilizers and plant hormone production, and suppressing diseases. The presence of PGPR in the plant rhizosphere is influenced by the water content of the soil. According to Van Loon (2007), PGPR promotes plant development through the production of plant hormones, improving nutrient absorption, or preventing the presence of pathogens. PGPR promotes plant growth through several mechanisms, including the secretion of Phyto hormones, fixing nitrogen, controlling ethylene production in roots, making nutrients like phosphate more soluble, producing siderophores, enhancing mycorrhizal function, and reducing the harmful effects of heavy metals. PGPR improves plant development in many ways, including germination rates, root development, productivity, hydraulic activity, chlorophyll content, Mg amount, N level, amount of protein, resistance to dryness, weight of shoots and roots, and delay of senescence of the leaves. Furthermore, it gives the plant resistance to diseases, which can be considered a kind of "biological control."

PGPR is a sustainable method for raising crop quality and production as it can improve root health and plant development in a variety of crop species. This system is well suited for sustainable agriculture and can enhance the production, importance, and production of agricultural goods that promote human health. As a result, it is very compatible with the practice of organic farming (Chandran et al., 2021) (Fig. 2).

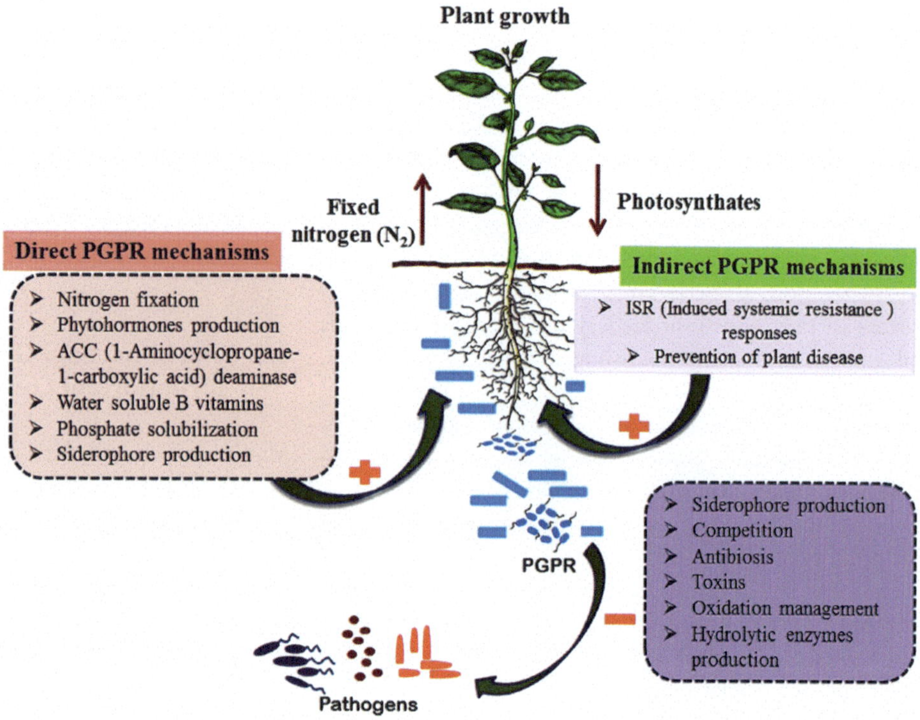

Fig. 2 The diagram illustrates the PGPR mechanism (*Source* Chandran et al., 2021)

4 Cyanobacteria

Cyanobacteria are autotrophic prokaryotes that carry out photosynthesis and are present in many habitats, such as lakes, ponds, springs, marshes, streams, and rivers. They play a vital role in maintaining soil health and have been recognized as natural biofertilizers in rice fields (Sahu et al., 2012). Unattached cyanobacteria or cyanobacteria that live in a mutually beneficial relationship with other organisms, such as *Nostoc linkia* and *Anabaena variabilis*, convert 4–6 teragrams of nitrogen gas into a usable form each year. Nitrogen-fixing cyanobacteria such as *Nostoc linkia, Anabaena variabilis, Aulosirafertilisima, Calothrix* sp., *Tolypothrix* sp., and *Scytonema* sp. have been employed to enhance rice production in different agro-ecological zones. Recently, researchers have conducted experiments to evaluate the application of A. variabilis, a type of algae, to different crops. In particular, they have studied the effects of using a mutant strain of A. variabilis that produces ammonia on the growth of wheat crops. The studies indicated that the wheat crops had enhanced growth (Manjunath et al., 2011). Cyanobacteria are essential for the global nitrogen cycle and the generation of phytohormones in both independent and symbiotic relationships. In rice and wheat rhizospheres, they synthesize indole acetic acid (IAA) and release extracellular polysaccharides, peptides, and lipids as they grow. In addition, they aid in the process of soil formation through the secretion of extracellular polysaccharides, peptides, and lipids. According to several accounts, cyanobacteria have the ability to thrive in saline soil, which enhances their fertility and inhibits the development of most plants, with the exception of halophytes.

Cyanobacteria, a group of photosynthetic microorganisms, are frequently employed as reliable indicators for assessing the presence of heavy metal pollutants in soil. A major factor contributing to heavy metal pollution in soil is the application of sewage sludge, which contains significant amounts of these toxic elements. Therefore, the utilization of cyanobacteria as bioindicators offers a valuable approach for monitoring and evaluating the extent of heavy metal contamination resulting from sewage sludge application in soil ecosystems. Nonetheless, the majority of experimental studies have consistently demonstrated an inverse relationship between nitrogenase activity or cyanobacteria and heavy metal concentration. It has been observed that cyanobacteria exhibit a high degree of sensitivity to experimental conditions, which may limit their suitability for reliable detection of metal contamination. Alternatively, it is recommended to assess the potential N2-fixation by conducting measurements under controlled laboratory conditions. Despite this fact, it is worth noting that cyanobacteria have been suggested as a suitable candidate for serving as an early warning system for detecting heavy metal contamination in the Swiss soil monitoring network. Cyanobacteria can be identified through the utilization of MPN techniques (Scherr et al., 2001) or by measuring nitrogenase activity with light energy. The measurement of nitrogenase activity is conducted by the acetylene reduction test, which generates ethylene as a byproduct. The quantification of ethylene can be conveniently accomplished via gas chromatography (Nielsen and Winding, 2002).

5 The Impact of Microorganisms on Managing the Environment

The global community faces a multitude of environmental issues, including pollution, waste production, land deterioration, and climate change resulting from human actions such as economic growth, urbanization, and forest destruction. Studies have demonstrated the efficacy of microorganisms in addressing these issues and recognizing them as viable instruments for attaining an environment that is viable. This section explores the diverse functions of microorganisms in efficiently addressing environmental issues.

5.1 Various Microorganisms Utilized as Bioremediation Agents

Industry-related activities, such as the application of fertilizers, pesticides, and herbicides, have a substantial impact on the contamination of both soil and aquatic habitats. The contaminants include nickel, lead, copper, zinc, cadmium, and arsenic. The presence of untreated contaminants in the wastewater generated by the agro-food industry is a significant environmental hazard (Hashim et al., 2022). According to Manjunath et al. (2011), the presence of nitrogen dioxide and sulfur dioxide in the atmosphere has a detrimental effect on plant development and agricultural productivity. Furthermore, ozone exposure may disrupt the photosynthetic cycle and metabolism in plants. Bioremediation is a highly efficient, environmentally sustainable, and financially viable approach for converting pollutants. Plants and microorganisms may be used for the purpose of biological remediation. Nevertheless, it is crucial to recognize that plants possess a longer life cycle and provide challenges in terms of their administration. Bioremediation may be conducted with bacteria, fungus, and algae. Microorganisms have the capacity to alleviate the impacts of heavy metals, enhance soil fertility, and promote plant growth, making them a more appropriate choice for remediation. Microorganisms metabolize poisonous chemicals, transforming them into water, CO_2, and other less hazardous molecules. These substances are then broken down by other bacteria via the process of mineralization (Khan et al., 2024b).

Immobilization refers to the process of transforming substances into a state where they are no longer accessible in the atmosphere. For example, we will analyze the process of converting nitrates into elemental nitrogen as outlined by Prasanna et al. (2016). Microorganisms are found everywhere in nature and have the ability to break down a variety of contaminants under unusual conditions. Their adaptability to various habitats enhances their effectiveness, since acidophiles thrive in acidic conditions, psychrophiles flourish in cold temperatures, and halophiles thrive in saltwater surroundings.

The use of heavy metal bioremediation is often employed, particularly in areas characterized by elevated pollution levels. Environmental remediation processes include bioleaching, a widely acknowledged method to effectively dissolve solid heavy metals present in sediments by employing acidophilic bacteria. This method has shown significant effectiveness in removing iron or sulfur impurities. The process of iron or sulfur oxidation involves the use of many bacteria, including *A. thiooxidans, Aspergillus sp., Mucor sp., Penicillium sp., Cladosporium sp., and Rhizopus sp* (Medfu Tarekegn et al., 2020). The process involves creating an acidic environment, which leads to the dissolution of the immobilized heavy metals and their transfer into a liquid solution.

Siderophores are produced by microorganisms in contaminated environments with low iron levels. Siderophores are chemical compounds that have the ability to bind to iron and are composed of hydroxamate, catecholate, and phenolate functional groups. Siderophores have the ability to bind to heavy metals, hence enhancing their solubility via the formation of complexes. Additionally, these compounds generate reactive oxygen species, which enhances their efficacy as agents for bioremediation of organic contaminants. Because cyanobacteria produce siderophores such as schizokinen, anachelin, and synechobactin, which can effectively break down complex chemicals like polythene, they are particularly effective in bioremediation. This makes them appropriate for a diverse array of activities (Ayilara & Babalola, 2023).

5.2 Exploring Various Microorganisms for Biodegradation

Biodegradation refers to the process of converting organic substances into microbial biomass, as a consequence of the discharge of water and CO_2 into the environment. Biomass-derived microbes (BSMs) possess a wide range of catabolic varieties and are capable of breaking down, altering, or accumulating different substances such as hydrocarbons, PCBs, PAHs, and radionuclides (Glick, 2010). Bio stimulant microorganisms (BSMs) such as PGPR and PGPF generate enzymes that can break down contaminants. These enzymes include peroxidases, dioxygenases, dehalogenases, nitrilases, and nitro reductases. PGPR bacteria are efficient rhizosphere bacteria that utilize the exudates of roots as a source of fuel and carbon and actively contribute to the breakdown of organic compounds by direct or co-metabolic processes. *Pseudomonas* and *Brevibacillus*, which promote the growth of plants, have been found to be effective in breaking down or eliminating petroleum-contaminated environments.

6 The Function of Soil Microbes in Mitigating Soil-Borne Plant Diseases

The microbiota is essential in determining the effectiveness of defensive methods, such as the addition of municipal waste compost to greenhouse conditions, in preventing soil-borne diseases. The regular combination of compost and wood with vegetables grown beneath plastic tunnels might gradually diminish the soil's ability to resist soil-borne illnesses. Biocontrol-based microbiota, primarily composed of bacteria, actinomycetes, and filamentous fungi, can initiate disease suppression mechanisms. Researchers have observed that the presence of Plant Growth-Promoting Rhizobacteria (PGPR) and non-pathogenic species of Fusarium and Pythium in disease-suppressive compost can positively influence plant growth and overall vegetative vigor. This phenomenon is attributed to the ability of these microorganisms to enhance the host's resistance or tolerance to various diseases (Hasan et al., 2024).

The growth and nutrient availability of plants can be improved through the presence of various microbiota in the soil, such as fluorescent pseudomonads, Bacillus, Paenibacillus, actinomycetes, and arbuscular mycorrhizal fungi (AMF). The microbes present in the rhizosphere exhibit the remarkable ability to not only inhibit the growth and proliferation of harmful pathogens but also enhance the absorption of essential nutrients by plants. Additionally, these beneficial microbes are known to synthesize and release substances that stimulate the growth and development of plants (De Corato, 2020). *Paenibacillus brasilensis*, a bacterium, can perform N_2 fixation, a process by which it changes atmospheric nitrogen into a usable form. Additionally, this bacterium is capable of producing two plant hormones, auxin and cytokinin. These hormones have a crucial role in different physiological processes in plants, particularly in development, and stress responses. *Paenibacillus brasilensis* possess the capacity to perform nitrogen fixation as well as generate auxin and cytokinin highlights its potential as a beneficial microorganism. The utilization of Bacillus, actinomycetes, and arbuscular mycorrhizal fungi (AMF) has been found to have a positive impact on the uptake of soluble phosphorus by plants (Deepa et al., 2010). Additionally, these microbes have been observed to stimulate alterations in the structure of the root system, leading to enhanced nutrient acquisition. Furthermore, they have been shown to alter the composition and dynamics of the rhizosphere and the soil environment surrounding the roots. Moreover, these microorganisms possess a vital role in maintaining a balance between root loss and the production of new biomass. The capacity of soil to suppress plant pathogens is contingent upon its ability to facilitate microbial growth and exhibit bioactivity.

7 The Importance of biotechnology in the Agricultural Sector

The use of agricultural technology throughout the twentieth century resulted in a significant increase in agricultural productivity, sometimes referred to as the green revolution. However, this progress came at the expense of negative ecological consequences, including global pollution, climate change, and the loss of biodiversity. The objective of biotechnology and microbiology studies is to enhance the effectiveness, yield, and environmental impact of crops by using genetically modified organisms (GMOs) and hybrid crops, thereby facilitating the creation of superior agricultural goods (Suman et al., 2022). The advancement of microbial biotechnology has made significant contributions to organic farming via the development and use of many applications, such as biofertilizers, biological pesticides, natural herbicides, bioinsecticides, and more. The use of microbial biotechnology mitigates the reliance on agricultural products in sustainable farming by controlling both living organism-related and non-living environment-related challenges. The advancement of microbial biotechnology has made significant contributions to organic farming via the development and use of many applications, such as biofertilizers, biological pesticides, natural herbicides, bioinsecticides, and more. The use of microbial biotechnology mitigates the reliance on agricultural products in sustainable farming by controlling both living organism-related and non-living environment-related challenges. The utilization of molecular biotechnology techniques has significantly enhanced the precision of plant breeding. These techniques encompass various tools such as DNA isolation, cloning of genes, and gene transfer between different species (Hart et al., 2018). Soil microorganisms play a vital role in ensuring the long-term viability of food production, particularly in the rhizosphere, where a diverse array of microbes coexist. Rhizosphere microorganisms in biotechnology play a crucial role in promoting agricultural sustainability by enhancing nitrogen fixation, boosting nutrient availability, controlling stress, and enhancing both the quality and quantity of crops. The use of microbial biotechnology has many advantages, such as less reliance on synthetic fertilizers, insecticides, and herbicides and the preservation of a natural ecosystem for subsequent generations. Additionally, it aids in the prevention of harmful contaminants and trash that have adverse effects on ecological systems, renewable materials. Microbial biotechnology provides a multitude of advantages for agricultural activities as a whole. The fields of biotechnology and microbiology are working together to improve the quality, production, and sustainability of crops by using genetically modified organisms (GMOs) and transgenic varieties of crops. Genetically altered crops exhibit many traits, such as regulated ripening, immunity to insects, resistance to fungi, resistance to viruses, and the ability to tolerate herbicides. These techniques enhance the longevity, durability, and overall quality of agricultural goods while also minimizing the need for insecticides and fungicides. In addition, they decrease the need for pesticides in several crops, such as tomatoes, potatoes, maize, rice, lettuce, coffee, the cabbage family, apples, and potatoes. Moreover, these crops have the potential to decrease the reliance on pesticides and enhance weed management (Suman et al., 2022).

Biotechnological intervention is crucial in this setting for identifying the microbiota. Recently, the combination of genetic material sequencing and increased computational ability has been widely utilized for identifying microbes in the microbiomes of various agroecology. This approach allows for the identification of microorganisms in any possible natural world without the need to separate and cultivate the microbes. Utilizing a wide range of information is very valuable for developing strategies to manipulate microbiomes to increase nutrient absorption, improve tolerance to both living and non-living stresses, and ultimately optimize agricultural yield. The combination of genome sequencing and bioinformatic resources has not only enabled fast detection of microbiomes. Additionally, it has it possible to annotate gene microbiomes quickly (Rebello et al., 2021). A next-generation sequencing (NGS), a newly developed technique, is already being used due to its significant benefits compared to the traditional Sanger sequencing approach. The improvement in this area of science has created rapid sequencing, that has been supplementary boosted the different results, which previously looked to be hard and tedious by standard approaches. The finalized sequencing results are of exceptional quality since they have been thoroughly characterized. These sequences provide very accurate genomic data and may be used for model organisms and agriculturally significant microbes (Khan et al., 2024b). Metagenomics investigations using SMRT (single-molecule real-time sequencing) have enabled the discovery and identification of microbiomes. This approach is inexpensive and does not need a preceding PCR step. Pyrosequencing is now used to analyze the taxonomic composition of microbiomes in soils by examining 16S/18S rDNA amplicons. Agricultural methods are globally used to achieve balanced ecological and commercial growth purposes. Agricultural microbiology is a specialized area of study that combines information from basic microbiology and the ecology of microbes with biotechnology for agriculture. Studying the frequent movement of microorganisms between plants, animals, and soil-based environments is essential for understanding the composition of microbiota in both organic and farmed habitats (Rebello et al., 2021).

8 Microbial Associations Within Various Layers of Soil

The composition of soil consists of a wide range of microorganisms, which can be categorized into seven groups: fungi, viruses, bacteria, nematodes, actinomycetes, blue-green algae, and protozoa. The presence of such microorganisms can be observed in both surface and subsurface soil. However, their abundance decreases as the soil depth increases, primarily due to the lack of organic materials on the earth's surface. Changes in the surroundings can lead to alterations in the composition of the soil's microbial community. The arrangement of microbes in soil varies among various geographic regions, with the rhizosphere, a narrow region directly influenced by plant roots, containing a higher concentration of bacteria compared to other portions. The rhizosphere is a constantly changing habitat where plant roots emit substances that promote greater numbers and functions of microorganisms in comparison to the adjacent soil. The abundance of microbes decreases with increasing soil depth as a result of reduced levels of organic matter. Environmental changes can lead to alterations in the soil's microbial ecology. The rhizosphere is a constantly changing microenvironment in which roots of plants discharge substances that encourage the development and function of larger populations of microbes (Sapkota, 2023).

Bacteria are abundant in soil, primarily located in the rhizosphere. In contrasting to bulk soil, they exhibit a greater size and possess an increased concentration of Gram-negative and denitrifying bacteria. The soil bacterial biomass exhibits the range of 300 and 3000 kg per hectare, with an estimated range of 107 to 109 bacteria per gram. Some examples of common bacterial genera are *Bacillus, Arthrobacter, Pseudomonas, Agrobacterium, Alcaligenes, Clostridium, Flavobacterium, Corynebacterium, Micrococcus, Xanthomonas, and Mycobacterium.* The rhizosphere is the predominant region where metabolically active bacteria are present, as there is an abundance of nutrients. The majority of bacteria have a preference for soils that are abundant in nutrients, have a pH that is either neutral or slightly alkaline, and possess a closely balanced carbon-to-nitrogen ratio. Fungi comprise the most prevalent living matter in the soil of various habitats, with a biomass ranging from 500 to 5000 kg per hectare. Fungi can be classified into many categories, such as decomposers, mutualists (mycorrhizal fungus), and pathogens. Some of the common fungi that can be found in soil are *Amanita,*

Tricholoma, Torrendia, Descomyces, Thelephora, Verticillium, Phytophthora, Rhizoctonia, and *Pythium.* Fungi exhibit a preference for soil with a low pH and a wide carbon-to-nitrogen ratio, which allows them to thrive in raw humus as well as moderate and mull soils, where they tend to be the dominant organisms (Tabassum et al., 2023). Mycorrhizal fungi primarily inhabit the vicinity of roots of plants, while other different categories are dispersed across the soil. Fungi have crucial functions in soil, including the cycling of nutrients, the regulation of water dynamics, and the prevention of diseases. These functions are vital for maintaining soil health and enhancing crop productivity. Mycorrhizal mutually beneficial relationships are formed with terrestrial plants, which improve the absorption of nutrient-rich minerals in return for carbon molecules produced by photosynthesis. Fungi also synthesize antibiotic substances such as penicillin and cyclosporin. Lichens, which are mutualistic partnerships between fungi and either algae or cyanobacteria, are essential organisms that colonize soil and have vital functions in the processes of soil development and maintenance. Actinomycetes play a crucial role in acidic soils because they have the capacity to withstand a broader spectrum of pH levels. Actinomycetes are fibrous bacteria that are primarily Gram-positive and are commonly found in soils with a pH ranging from neutral to alkaline. These organisms create colonies or large networks of fungal threads, known as mycelia, that can detach and transform into rod- or coccoid-shaped structures. The highest concentration of actinomycetes is found in the topmost layer of soil, although they are present in all layers of soil. They proliferate as filamentous fungi, which leads to the characteristic earthy aroma of recently disturbed soil. Streptomyces is the predominant species, with *Micromonospora, Thermoactinomycetes, and Nocardia following in abundance. Actinomycetes* such as *Microbispora, Microtetraspora, Amycolaptosis, Actinomadura, and Saccharothrix* are infrequently seen in arid soil. Approximately 90% of Actinomycetes genera have been extracted from soil, where they develop branched hyphae that fragment into spores. Actinomycetes are organisms that thrive in moderate temperatures and are highly susceptible to acidity and low pH levels (Tabassum & Hashim, 2023). Additionally, certain types of actinomycetes can be harmful, leading to the development of plant diseases. Actinomycetes serve an essential role in soil cultivation by contributing to the carbon cycle through photosynthesis and decomposition. They form symbiotic relationships with non-leguminous plants, performing nitrogen fixation, which subsequently becomes accessible to both the hosts and neighboring plants. Actinomycetes are the predominant category of soil microbes responsible for breaking down decaying matter by producing hydrolytic enzymes. They enhance plant development, enhance the accessibility of nutrients, produce growth promoters, and inhibit pathogenic microbes. Additionally, these characteristics may be employed as biological control instruments.

Protozoa, such as amoeba, ciliates, and flagellates, can be found in several types of soil, including agricultural, grassland, forest, and freshwater bottom sediment. Their occurrence is controlled by the existence of root system of plants and the organic composition of the soil. The distribution of protozoa is influenced by the characteristics of the soil, such as its structure and texture. Mastigophorans (flagellates) are more prevalent in drier soils, while ciliophorans (ciliates) are more common in humid soils (Geisen et al., 2018). Flagellates and amoebae are the main types of microorganisms found in cultivated soil and clay soil, whereas coarse soil contains larger flagellates, testaceans, and ciliates. Certain protozoans have the capacity in order to form symbiotic relationships with other microorganisms. Protozoans have a vital role in terrestrial ecosystems via consuming bacteria and converting organic soil nitrogen into ammonium. Furthermore, they also possess the capacity to evaluate and track alterations in soil conditions, serving as markers of biological activity. Protozoa enhance plant biomass regardless of nutrient composition, and numerous species consume bacteria, hence promoting nutrient cycling and energy transfer across microbes, animals, and plants (Geisen et al., 2018; Tabassum et al., 2023).

Cyanobacteria are photosynthetic bacteria that inhabit damp soils, predominantly in the immediate vicinity of the roots of plants. They possess the ability to exist independently in the soil or establish mutually beneficial partnerships with fungi that generate lichens. Some examples of cyanobacteria are *Nostoc, Prochlorothrix, Anabaena,* and *Nodularia.* These creatures flourish on a diverse array of soils, encompassing both the surface and subsurface. These organisms can withstand moderate temperatures and are affected by acidity and low pH levels. They are also one of the earliest communities of microorganisms to inhabit terrestrial environments. Cyanobacteria have crucial functions in soil as they do nitrogen and carbon fixation through exopolysaccharides, which enhances soil fertility and improves its water holding capacity. In addition, their filamentous form enhances soil physical structure, porosity, and moisture retention. They generate mucilaginous compounds and secrete phytohormones, vitamins, amino acids, and secondary metabolites. Several species of cyanobacteria have the capability to convert atmospheric N_2 into a usable form by using specialized cells known as heterocyst (Geisen et al., 2018). Cyanobacteria have the potential to function as a source of N2-biofertilizer in fields, offering a substitute for natural and chemical-based fertilizers. Nevertheless, the absence of a substantial cyanobacteria population can lead to oxygen depletion, which in turn impacts the community of bacteria.

Viruses are genetic entities that have the ability to replicate autonomously from a cell's chromosomes, although they are reliant on cells for their replication. Nanoparticles are diminutive compared to bacteria, varying in diameter from 20 to 30 nm (Tabassum et al., 2023). Viruses are parasitic organisms that are dependent on bacteria, fungi, insects, plants, and mammals for survival. They are specifically adapted to live in soil environments. Bacterial phages are the most widely distributed organisms in the environment, surpassing the quantity of cell-based organisms found in aquatic and soil environments (Geisen et al., 2018). The viral content in soil is estimated to be 109 virus particles per gram of dry weight. Soil viruses can be classified into two main groups. The initial category comprises small, spherical entities that bear resemblance to bacteriophages carrying RNA with a single strand that come from the Leviviridae family or specific viruses from plants. The second group comprises larger spherical viruses that bear resemblance to those comprising DNA with two strands and come from the families Partitiviridae, Chrysoviridae, and Totiviridae. Various viruses belonging to the *Myoviridae, Siphoviridae*, and *Podoviridae* families have been discovered on the surface sands of the Sahara desert. Soils serve as reservoirs for viruses, yet some viruses are capable of easily moving across different habitats, indicating that soils are not completely static (Geisen et al., 2018; Tabassum et al., 2023).

Nematodes are diminutive invertebrates characterized by sleek, unsegmented anatomies, often measuring fifty micrometers in diameter and one millimeter in length. They inhabit the uppermost layer of soil, where they consume living microorganisms found on the surfaces of the soil. These organisms can exist independently or as parasites and can be categorized into four groups: bacterivores, fungivores, proto zoophages, and omnivores. Soil nematodes are essential for maintaining the community of microbes and ensuring an adequate supply of nitrogen for plants (Tabassum et al., 2023). Certain autonomous nematodes possess the potential to transform chemical molecules into minerals, which helps in the process of biogeochemical cycles and improves the fertility of the soil. Certain nematodes engage in parasitism or predation to obtain nourishment from pests. The locomotion of nematodes inside the soil enhances soil porosity, hence promoting ecological equilibrium. In general, nematodes have a vital function in preserving the microbial ecosystem of soil and guaranteeing the accessibility of N_2 for plants.

9 Conclusion

The microbiome exerts a significant impact on sustainable agriculture by determining soil health, plant composition and yield, and the sustainability of agricultural practices as a whole. The diversity and abundance of plants significantly impact their composition, yield, and resilience to climate variations. In this context, biotechnological intervention is essential for microbiota identification. In recent times, there has been extensive application of the combination of genetic material sequencing and enhanced computational capabilities in the identification of microorganisms within the microbiomes of diverse agroecology. Engineering techniques to modify the soil microbiome have the potential to significantly improve agriculture through the mitigation of plant illnesses and the enhancement of plant resilience to the adverse effects of climate change. Soil microbes are of the utmost significance in the provision of ecosystem services and have the potential to alleviate the detrimental effects that human activities have on soil ecosystems. These organisms perform essential functions and have the ability to decompose hazardous substances such as heavy metals and pesticides. Soil-dwelling mycorrhizal fungi, nitrogen-fixing bacteria, and additional microorganisms associated with roots comprise the rhizosphere. The rhizosphere exhibits the capability of degrading pesticides, while viruses, bacteria, and nematodes are potential detrimental species. The study emphasizes the critical role that rhizosphere microorganisms play in mitigating these concerns.

References

Andreote, F. D., Gumiere, T., & Durrer, A. (2014). A Exploring interactions of plant microbiomes. *Science in Agriculture, 71*, 528–539.

Aseri, G. K., Jain, N., Panwar, J., Rao, A. V., & Meghwal, P. R. (2008). Biofertilizers improve plant growth, fruit yield, nutrition, metabolism and rhizosphere enzyme activities of pomegranate (Punica granatum L.) in Indian Thar Desert. *Scientific Horticulture, 117*(2), 130–135.

Ayilara, M. S., & Babalola, O. O. (2023). Bioremediation of environmental wastes: The role of microorganisms. *Frontiers in Agronomy, 5*, 1183691. https://doi.org/10.3389/fagro.2023.1183691

Bar On, Y. M., Phillips, R., & Milo, R. (2018). The biomass distribution on Earth. *Proceedings of the National Academy of Sciences of the United States of America, 115*, 6506–6511. https://doi.org/10.1073/pnas.1711842115

Bunea, A., Pacurar, I., Dulf, V., et al. (2020). Study concerning the soil resources used in agriculture and forestry worldwide. *ProEnvironment, 13*, 39–43.

Chandran, H., Meena, M., & Swapnil, P. (2021). Plant growth-promoting rhizobacteria as a green alternative for sustainable agriculture. *Sustainability, 13*, 10986. https://doi.org/10.3390/su131910986

Compant, S., Samad, A., Faist, H., Sessitsch, A. (2019). A review on the plant microbiome: Ecology, functions, and emerging trends in microbial application. *Journal of Advanced Research, 19*, 29–37. https://doi.org/10.1016/j.jare.2019.03.004

De Corato, U. (2020). Soil microbiota manipulation and its role in suppressing soil-borne plant pathogens in organic farming systems under the light of microbiome-assisted strategies. *Chemical and Biological Technologies in Agriculture, 7*(1). https://doi.org/10.1186/s40538-020-00183-7

De-Bruijn, F. J. (2015). Biological nitrogen fixation. In Lugtenberg (Ed). *Principles of plant-microbe interactions* (pp. 215–224). Springer International Publishing Switzerland.

Deepa, C., Dastager, S. G., & Pandey, A. (2010). Plant growth-promoting activity in newly isolated Bacillus thioparus (NII-0902) from Western ghat forest, India. *World Journal of Microbiology and Biotechnology, 26*(12), 2277–2283.

Dignam, B. E., O'Callaghan, M., Condron, L. M., Kowalchuk, G. A., Van Nostrand, J. D., Zhou, J., & Wakelin, S. A. (2018). Effect of land use and soil organic matter quality on the structure and function of microbial communities in pastoral soils: Implications for disease suppression. *PLoS ONE, 13*(5), Article e0196581.

Geisen, S., Mitchell, E. A. D., Adl, S., Bonkowski, M., Dunthorn, M., Ekelund, F., Fernández, L. D., Jousset, A., Krashevska, V., Singer, D., Spiegel, F. W., Walochnik, J., & Lara, E. (2018). Soil protists: a fertile frontier in soil biology research. *FEMS Microbiology Reviews, 42*(3), 293–323. https://doi.org/10.1093/femsre/fuy006

Girija Veni, V., Srinivasarao, C., Sammi Reddy, K., et al. (2020). Soil health and climate change. In M. N. V. Prasad & M. Pietrzykowski (Eds.), *Climate change and soil interactions* (pp. 751–767). Elsevier.

Glick, B. R. (2010). Using soil bacteria to facilitate phytoremediation. *Biotechnology Advances, 2*, 367–374.

Hart, M. M., Antunes, P. M., Chaudhary, V. B., et al. (2018). Fungal inoculants in the field: Is the reward greater than the risk? *Functional Ecology, 32*(1), 126–135.

Hasan, A., Tabassum, B., Hashim, M., & Khan, N. (2024). Role of Plant growth promoting rhizobacteria (PGPR) as a plant growth enhancer for sustainable agriculture: A review. *Bacteria, 3*, 59–75. https://doi.org/10.3390/bacteria302000

Hashem, A., Tabassum, B., & Abd_Allah, E. F. (2019). Bacillus subtilis: A plant-growth promoting rhizobacterium that also impacts biotic stress. *Saudi Journal of Biological Sciences, 26*(6), 1291–1297.

Hashim, M., Arif, H., Tabassum, B., Arif, A., Rehman, A. A., Rehman, S., Khanam, R., Khan, B., Hussain, A., Barnawi, J., Tayeb, F. J., Algehainy, N., Altemani, F. H., Mir, R., Almutairi, F. M., Ullah, M. F., Elfaki, I., & Ajmal, M. R. (2022). Protective effect of catharanthus roseus extract on cadmium-induced toxicity in albino rats: a putative mechanism of detoxification. *Metabolites, 12*(11), 1059. https://doi.org/10.3390/metabo12111059

Hunter, P. (2016). Plant microbiomes and sustainable agriculture. *EMBO Report, 17*, 1696–1699. https://doi.org/10.15252/embr.201643476

Jansson, J., McClure, R., & Egbert, R. (2023). Soil microbiome engineering for sustainability in a changing environment. *Nature Biotechnology, 41*. https://doi.org/10.1038/s41587-023-01932-3

Johns, N. I., Blazejewski, T., Gomes, A. L., & Wang, H. H. (2016). Principles for designing synthetic microbial communities. *Current Opinion in Microbiology, 31*, 146–153. https://doi.org/10.1016/j.mib.2016.03.010

Kacem, M., Kazouz, F., Merabet, C., Rezki, M., de Lajudie, P., & Bekki, A. (2009). Antimicrobial activity of Rhizobium sp. strains against Pseudomonas savastanoi, the agent responsible for the olive knot disease in Algeria. *Grasas Y Aceites, 60*(2), 139e146.

Khan, M., Tabassum, B., Hashim, M., & Khan, T. (2024a). Exploring the utilization of plant biomass in addressing soil contamination by potential toxic metals through phytoremediation. *Microbiome-Assisted Bioremediation*, 429–451. https://doi.org/10.1016/b978-0-443-21911-5.00018-0

Khan, N., Tabassum, B., Hashim, M., & Hasan, A. (2024b). Effect of bio-sorptive removal of heavy metals from hydroponic solution: A review. In N. Kumar (Eds.) *Hydroponics and environmental bioremediation.* Springer Water. Springer. https://doi.org/10.1007/978-3-031-53258-0_12

Kumar, A., Sharma, S., Mishra, S., & Dames, J. F. (2015). Arbuscular mycorrhizal inoculation improves growth and antioxidative response of Jatropha curcas (L.) under Na_2SO_4 salt stress. *Plant Biosystem, 149*, 260–269. https://doi.org/10.1080/11263504.2013.845268

Kumar, I., Mondal, M., Gurusamy, R., Balakrishnan, S., & Natarajan, S. (2019). Plant-microbiome interaction and the effects of biotic and abiotic components in agroecosystem. *Microbial Interventions in Agriculture and Environment, 2*, 517–546. https://doi.org/10.1007/978-981-13-8383-0_18

Lim, J. H., & Kim, S. D. (2013). Induction of drought stress resistance by multi-functional PGPR Bacillus licheniformis K11 in pepper. *The Plant Pathology Journal, 29*, 201–208. https://doi.org/10.5423/PPJ.SI.02.2013.0021

Manjunath, M., Prasanna, R., Shama, P., Nain, L., & Singh, R. (2011). Developing PGPR consortia using novel genera Providencia and Alcaligenes along with cyanobacteria for wheat. *Archives of Agronomy*

Medfu Tarekegn, M., Zewdu Salilih, F., & Ishetu, A. I. (2020). Microbes used as a tool for bioremediation of heavy metal from the environment. *Cogent Food & Agriculture, 6*(1), 1783174. https://doi.org/10.1080/23311932.2020.1783174

Mehta, M. P., & Baross, J. A. (2006). Nitrogen fixation at 92 C by a hydrothermal vent archaeon. *Science, 314*, 1783–1786.

Moussa, T. A. A., Almaghrabi, O. A., & Abdel-Moneim, T. S. (2013). Biological control of the wheat root rot caused by Fusarium graminearum using some PGPR strains in Saudi Arabia. *The Annals of Applied Biology, 163*, 72–81.

Nielsen, M. N., & Winding, A. (2002). *Microorganisms as indicators of soil health*, Technical Report No. 388. National Environmental Research Institute

Nihorimbere, V., Ongena, M., Smargiassi, M., & Thonart, P. (2011). Beneficial effect of the rhizosphere microbial community for plant growth and health. *Biotechnologie, Agronomie, Societe Et Environnement, 2*, 327–337.

Prasanna, R., Nain, L., Rana, A., Shivay, Y. S. (2016). Biofortification with microorganisms: Present status and future challenges. *Biofortification Food Crops*, 249–62. https://doi.org/10.1007/978-81-322-2716-8_19

Rebello, S., Nathan, V. K., Sindhu, R., Binod, P., Awasthi, M. K., & Pandey, A. (2021). Bioengineered microbes for soil health restoration: Present status and future. *Bioengineered, 12*(2), 12839–12853. https://doi.org/10.1080/21655979.2021.2004645

Ryu, M.-H., Zhang, J., Toth, T., et al. (2020). Control of nitrogen fixation in bacteria that associate with cereals. *Nature Microbiology, 5*(2), 314–330.

Sahu, D., Riyadarshani, P., & Rath, B. (2012). Cyanobacteria: As potential biofertilizer. *Journal of Microbiology, 1*, 20–26.

Sapkota, A. (2023). *Microorganisms found in soil with effects and examples*. Microbe Notes. https://microbenotes.com/microorganisms-in-soil

Scherr, C., Fliessbach, A., & Mäder, P. (2001). Photoautotrophe Boden mikroorganismen als Bioindikatoren für Schwermetalle. *VBB-Bulletin, 5*, 7–9.

Singh, D. P., Singh, H. B., Prabha, R. (2016). Microbial inoculants in sustainable agricultural productivity. In *Soil microbes: The invisible managers of soil fertility* (pp. 1–16). https://doi.org/10.1007/978-81-322-2644-4

Suman, J., Rakshit, A., Ogireddy, S. D., Singh, S., Gupta, C., & Chandrakala, J. (2022). Microbiome as a key player in sustainable agriculture and human health. *Frontiers in Soil Science, 2*, Article 821589. https://doi.org/10.3389/fsoil.2022.821589

Tabassum, B., Hashim, M., & Dhingra, J. M. (2023). Potency of three cruciferous plants extracts as agro-phyto-remidiator against root knot nematode Meloidogyne spp. in Daucus carota (Carrot) under climate stress conditions. In J. A. Parray (Ed.), *Climate change and microbiome dynamics*. Climate Change Management. Springer. https://doi.org/10.1007/978-3-031-21079-2_17

Tabassum, B., & Hashim, M. (2023). Beneath the soil: Exploring nematode fauna dynamics and crop infection patterns in Bareilly, Uttar Pradesh, India. *Micro Environer, 3*(2), 14–24.

Tardif, S., Yergeau, É., Tremblay, J., Legendre, P., Whyte, L. G., & Greer, C. W. (2016). The willow microbiome is influenced by soil petroleum-hydrocarbon concentration with plant compartment-specific effects. *Frontiers in Microbiology, 7*, 1363. https://doi.org/10.3389/fmicb.2016.01363

Thakur, P. (2012). *Screening of plastic degrading bacteria from dumped soil area* [Doctoral dissertation].

Timmusk, S., Islam, A., El Abd, D., Lucian, C., Tanilas, T., Kännaste, A., et al. (2014). Drought-tolerance of wheat improved by rhizosphere bacteria from harsh environments: Enhanced biomass production and reduced emissions of stress volatiles. *PLoS ONE, 9*, Article e96086. https://doi.org/10.1371/journal.pone.0096086

Trivedi, P., Delgado-Baquerizo, M., Trivedi, C., Hamonts, K., Anderson, I. C., & Singh, B. K. (2017). Keystone microbial taxa regulate the invasion of a fungal pathogen in agro-ecosystems. *Soil Biology and Biochemistry, 111*, 10. https://doi.org/10.1016/j.soilbio.2017.03.013

Van Loon, L. C. (2007). Plant responses to plant growth promoting bacteria. *European Journal of Plant Pathology, 119*, 243–254.

Velazquez, E., Carro, L., & Flores Felix, J. D. (2017). The legume nodule microbiome: A source of plant growth-promoting Bacteria. *Probiotics Plant Health*, 41–70. https://doi.org/10.1007/978-981-10-3473-2_3

Wendimu, A., Yoseph, T., & Ayalew, T. (2023). Ditching phosphatic fertilizers for phosphate-solubilizing biofertilizers: A step towards sustainable agriculture and environmental health. *Sustainability., 15*, 1713. https://doi.org/10.3390/su15021713

Zolla, G., Bakker, M. G., Badri, D. V., Chaparro, J. M., Sheflin, A. M., et al. (2013). Understanding root-microbiome interactions. In F. J. de Bruijn (Ed.). *Molecular microbial ecology of the rhzosphere* (vol. 2, pp. 745–754). Wiley Blackwell.

Climate Change: Impacts on Forest Cover

Partha Chandra Mondal, Arkadeb Mukhopadhyay, Madhu Tippannanavar, Shreosi Biswas, and Shweta Meshram

Abstract

Climate change, driven by greenhouse gas emissions, significantly impacts forest ecosystems, which cover 31% of the world's land and are vital for biodiversity, water regulation, and carbon sequestration. Rising temperatures, altered rainfall patterns, extreme weather events, and elevated CO_2 levels disrupt forest health, composition, and distribution. These changes lead to ecological impacts like phenological shifts, species range shifts, increased pest infestations, and biodiversity loss, altering ecosystem services. Temperature increases cause earlier leaf-out and shift species ranges, affecting forest composition. Altered precipitation patterns induce water stress and flooding, impacting tree health and soil stability. Extreme weather events like storms and wildfires cause significant physical damage. Elevated CO_2 levels enhance photosynthesis but shift competitive dynamics among species. Socio-economically, communities relying on forests face disruptions in livelihoods and the forestry industry, leading to economic risks. The cultural and recreational values of forests are also threatened. Mitigation and adaptation strategies include sustainable forestry practices, reforestation, conservation efforts, effective policy integration, and financial incentives like carbon credits. This chapter discusses the research and monitoring done so far to enhance forest resilience and protect biodiversity through these strategies can mitigate the adverse effects of climate change on forests, ensuring they continue to provide essential services and support communities dependent on them.

Keywords

Climate change · Forests · Biodiversity · Mitigation · Adaptation · Sustainability · Resilience

P. C. Mondal · S. Biswas
Division of Agricultural Chemicals, ICAR-Indian Agricultural Research Institute, New Delhi, India

A. Mukhopadhyay
R&D- Formulation Department, Insecticides (India) Limited, Rajasthan, India

M. Tippannanavar
AINP On Pesticide Residues, ICAR-Indian Agricultural Research Institute, New Delhi, India

S. Meshram (✉)
School of Agriculture, Lovely Professional University, Jalandhar, Punjab, India
e-mail: shweta.26662@lpu.co.in

1 Introduction

Climate change, driven by increased greenhouse gas emissions from human activities, leads to significant alterations in temperature, precipitation patterns, and atmospheric conditions (Adak et al., 2023). IPCC defines "Climate change as a change in the state of the climate that can be identified (e.g., by using statistical tests) by changes in the mean and/or the variability of its properties and that persists for an extended period, typically decades or longer. Climate change may be due to natural internal processes or external forcing such as modulations of the solar cycles, volcanic eruptions, and persistent anthropogenic changes in the composition of the atmosphere or in land use" (IPCC, 2018). These changes have far-reaching implications for natural ecosystems, particularly forests, which are essential components of the Earth's environmental health. Forests, which cover approximately 31% of the world's land area, provide critical ecological services such as maintaining biodiversity, regulating water cycles, and sequestering carbon dioxide (Nias & Mooney, 2013). They act as carbon sinks, absorbing more carbon than they emit, thus playing a pivotal role in mitigating climate change. However, the accelerated pace of climate change threatens the stability and functionality of these ecosystems. Rising

J. A. Parray (ed.), *Soil and Land Use Change*, Sustainable Landscape Planning and Natural Resources Management,
https://doi.org/10.1007/978-3-031-93075-1_11

temperatures, altered rainfall patterns, increased frequency of extreme weather events, and elevated CO_2 levels all contribute to changes in forest health, composition, and distribution (Leon et al., 2022).

The impacts of climate change on forest ecosystems are complex and multifaceted, affecting both ecological and socio-economic dimensions (De Frenne et al., 2021). Ecologically, forests face increased risks of pest infestations, disease outbreaks, and invasive species, all of which can lead to declines in tree health and biodiversity loss (Lawrence et al., 2022). Socio-economically, communities that depend on forests for their livelihoods, such as through timber production, non-timber forest products, and tourism, may experience significant disruptions (Prochazka et al., 2023). Furthermore, changes in forest cover can alter ecosystem services, impacting water quality and availability, soil stability, and carbon sequestration capabilities (Balthazar et al., 2015). This chapter explores the various mechanisms through which climate change influences forests, including temperature changes, precipitation variability, and extreme weather events. It also examines the resultant ecological consequences, such as species range shifts and increased fire risk, and socio-economic effects, such as livelihood disruptions and economic losses. Finally, the chapter discusses strategies for mitigating these impacts and adapting forest management practices to enhance resilience and sustainability in the face of a changing climate.

2 The Role of Forests in Global Climate Regulation

Forests play a crucial role in global climate regulation by acting as major carbon sinks, absorbing approximately one-third of the carbon dioxide emitted from fossil fuel combustion annually (Artaxo et al., 2022). This function not only mitigates climate change by reducing atmospheric CO_2 levels but also influences global weather patterns and climate systems. Healthy forests contribute to the stability of the hydrological cycle, influencing precipitation and water availability far beyond their immediate surroundings (Lykhovyd, 2021). The loss or degradation of forest cover can lead to a feedback loop of increased carbon emissions and reduced carbon absorption capacity, exacerbating climate change impacts (Hurteau, 2021). Moreover, forests are vital for maintaining air quality and supporting a diversity of species, including many that are critical for pollination and other ecosystem services essential to human agriculture and food security (Lukina et al., 2021). Recognizing the intrinsic value of forests and their multifaceted contributions to environmental and human health is fundamental to fostering a sustainable and resilient future. Addressing deforestation and forest degradation through integrated landscape management and cross-sectoral policies is imperative for safeguarding these essential ecosystems in the face of a rapidly changing climate (Morgan et al., 2021).

In addition to their critical role in carbon sequestration, forests also serve as vital reservoirs of biodiversity, hosting more than 80% of terrestrial species (FAO, 2020). This rich biodiversity not only underpins the resilience of ecosystems but also provides invaluable resources for medicine, agriculture, and industry. Furthermore, forests play a crucial role in regulating the Earth's temperature by influencing albedo—the reflection of solar radiation. Areas with dense forest cover tend to have lower albedo, absorbing more heat and contributing to the moderation of local and global temperatures (Bonan, 2008). Forest ecosystems also offer cultural, recreational, and spiritual benefits, enhancing the well-being of millions of people worldwide. However, the increasing pressure from deforestation, driven by agricultural expansion, logging, and infrastructure development, threatens to undermine these vital functions. Protecting and restoring forests through reforestation, afforestation, and sustainable management practices are essential strategies in mitigating climate change, preserving biodiversity, and ensuring the continued provision of ecosystem services critical for life on Earth.

3 Mechanisms of Climate Change Impact on Forests

The diverse effects of climate change on forest ecosystems have been represented in Fig. 1 and outlined in detail in the following subsections:

3.1 Temperature Changes

The increase in global temperatures, one of the most direct effects of climate change, has a cascade of effects on forest ecosystems. Trees and other forest vegetation have specific temperature ranges within which they thrive. As temperatures rise:

(i) Phenological Shifts: Trees may experience altered timing of life cycle events such as leaf-out, flowering, and fruiting. For instance, earlier spring warming can lead to premature leaf-out, making trees vulnerable to late frost events (Albrich et al., 2020).
(ii) Range Shifts: Species may migrate poleward or to higher elevations in search of suitable temperatures. However, not all species can move at the pace required, leading to potential mismatches in forest composition and ecosystem function (Carlson et al., 2022).

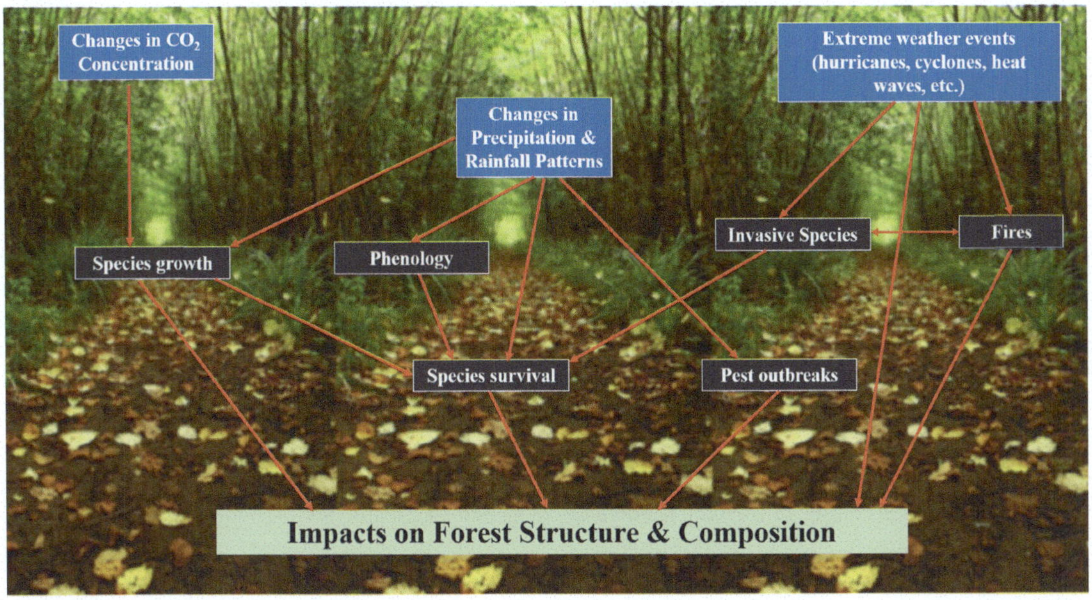

Fig. 1 Representative effects of climate change on forest ecosystems (Modified from Hergarten et al., 2013)

(iii) Increased Metabolism: Higher temperatures can accelerate tree metabolism and respiration rates, potentially increasing growth rates but also leading to higher water and nutrient demands.

3.2 Precipitation Patterns

Changes in precipitation, including altered frequency and intensity of rainfall and increased occurrence of droughts, significantly affect forest health and distribution (Felton et al., 2021):

(i) Water Stress: Prolonged drought conditions can lead to water stress, reducing tree growth, increasing mortality rates, and making forests more susceptible to pests and diseases.
(ii) Flooding: Increased rainfall intensity can cause flooding, leading to soil erosion, root damage, and altered nutrient dynamics in forest ecosystems.
(iii) Snowmelt and Water Availability: Changes in snowpack and the timing of snowmelt affect water availability in many forested regions, particularly in mountainous areas.

Changes in precipitation patterns not only impact the health and distribution of forests but also have cascading effects on the broader ecosystem services that forests provide. For example, altered rainfall regimes can disrupt the hydrological cycle, affecting groundwater recharge, river flow, and wetland ecosystems that are closely linked to forested areas (McLaughlin et al., 2020). This, in turn, can impact agriculture, drinking water supplies, and aquatic biodiversity. Moreover, extreme weather events, such as hurricanes and storms, which are often accompanied by heavy rainfall, can cause significant physical damage to forests, including deforestation and the destruction of canopy layers, which are critical for protecting the soil and maintaining forest microclimates (Uriarte et al., 2019). Additionally, the increased frequency of droughts and erratic precipitation can exacerbate the risk of wildfires, further threatening forest ecosystems and the services they provide to human populations. Effective forest management strategies must therefore consider the impacts of changing precipitation patterns to mitigate these risks and promote forest resilience in the face of climate change.

3.3 Extreme Weather Events

Climate change is linked to the increased frequency and severity of extreme weather events such as hurricanes, cyclones, and heat waves (Schuldt et al., 2020). These events can cause:

(i) **Physical Damage**: Storms can result in significant physical damage to trees through windthrow, breakage, and uprooting.
(ii) **Fire Regimes**: Higher temperatures and prolonged dry periods increase the risk and intensity of wildfires, which can decimate forest areas and alter ecosystem dynamics for decades.

3.4 CO_2 Concentration

Elevated atmospheric CO_2 levels, while generally beneficial for plant growth due to enhanced photosynthesis (a phenomenon known as CO_2 fertilization), can have complex effects on forests (McDowell et al., 2020):

(i) Altered Growth Patterns: Increased CO_2 can lead to faster growth rates in some tree species, but this is often limited by other factors such as nutrient and water availability.
(ii) Changes in Species Composition: Differential responses to CO_2 among species can shift competitive dynamics, potentially altering forest composition and biodiversity.

Elevated CO_2 levels not only influence growth patterns and species composition but also affect the overall health and resilience of forest ecosystems. While CO_2 fertilization may initially boost growth, this effect is often temporary and may lead to a phenomenon known as "carbon saturation," where trees become less responsive to increased CO_2 over time (Baron & Bélanger, 2020). Moreover, the accelerated growth spurred by higher CO_2 levels can lead to quicker depletion of essential nutrients in the soil, resulting in nutrient imbalances that could weaken trees and reduce their resistance to pests, diseases, and environmental stressors (Dotaniya, 2015). Additionally, the uneven response of different species to elevated CO_2 can cause shifts in forest structure, potentially favoring invasive or fast-growing species at the expense of slower-growing, native ones, thereby altering ecosystem function and stability (Xiao et al., 2021). Furthermore, increased CO_2 can influence the timing of phenological events such as flowering and fruiting, which may disrupt the interactions between plants and the animals that rely on them, with cascading effects on forest biodiversity and ecosystem services.

3.5 Ecological Impacts

(i) Biodiversity Loss: Climate change can lead to the loss of biodiversity within forest ecosystems through habitat alteration, where changing climate conditions render habitats unsuitable for many species, leading to declines or local extinctions. Additionally, species mismatches, such as phenological mismatches between trees and their pollinators or seed dispersers, can disrupt ecosystem functions. Furthermore, changing climates can facilitate the spread of invasive species that outcompete native flora and fauna, further threatening biodiversity (Weiskopf et al., 2020).
(ii) Altered Ecosystem Services: Forests provide a multitude of ecosystem services, including carbon sequestration, water regulation, soil protection, and recreational opportunities. Climate change can impair these services through reduced carbon sequestration, as stressed or dying trees sequester less carbon, diminishing the forest's role in mitigating climate change (Hasan et al., 2020). Additionally, changes in forest cover and health can disrupt local and regional water cycles, affecting water availability and quality. Furthermore, increased erosion and changes in soil composition due to altered forest cover can lead to soil degradation, reducing soil fertility and stability.
(iii) Forest Health and Productivity: Climate change can directly impact forest health and productivity by increasing pest and disease outbreaks, as warmer temperatures and stressed trees create favorable conditions for pests and pathogens, leading to widespread forest decline (Hartmann et al., 2022). Additionally, while some trees may grow faster under elevated CO_2 and warmer temperatures, this is often offset by increased mortality and other stress factors.

3.6 Socio-Economic Consequences

Forests are critical to the livelihoods of millions of people worldwide, particularly in rural and indigenous communities. Climate-induced changes in forest cover can disrupt livelihoods, as reduced forest productivity and biodiversity impact forest-based livelihoods, including agriculture, forestry, and non-timber forest products. Additionally, the forestry industry, which relies on predictable forest growth and health, faces significant economic risks from climate change, including reduced timber yields and increased management costs (Ela et al., 2021).

3.7 Cultural Impacts

Forests hold immense cultural significance for many communities, providing spiritual, recreational, and aesthetic values. Climate change threatens these cultural values through the loss of sacred sites, as changes in forest landscapes can affect sites of cultural and spiritual importance. Additionally, declines in forest health and biodiversity can reduce recreational opportunities, diminishing the recreational and aesthetic value of forest areas (Sesana et al., 2021).

4 Mitigation and Adaptation Strategies

The effects are highly evident in various environmental compartments. Consequently, multiple strategies, both individually and in combination, should be implemented to mitigate these adverse impacts. The impacts of climate change on forest ecosystems and corresponding adaptation and mitigation strategies have been summarized in Table 1 and outlined in detail in the following subsections.

4.1 Forest Management Practices

Adapting forest management practices to mitigate the impacts of climate change involves (Ontl et al., 2020):

(i) Sustainable Forestry: Implementing sustainable forestry practices that enhance forest resilience, such as selective logging, mixed-species plantations, and reduced impact logging.

Table 1 Impacts of climate change on forest ecosystems and corresponding adaptation and mitigation strategies

Climate change impact	Effects on forest ecosystems	Adaptation strategies	Mitigation strategies	References
Temperature increase	Shifts in species distribution and phenology	Promote climate-resilient species and genetic diversity	Enhance carbon sequestration through afforestation and reforestation	Adak et al. (2023)
	Increased vulnerability to pests and diseases	Implement adaptive forest management techniques	Foster sustainable forest management practices	
Altered precipitation patterns	Increased frequency of droughts and floods	Develop drought and flood-resistant species	Enhance water management practices	Felton et al. (2021)
	Changes in soil moisture and nutrient availability	Use soil conservation and water harvesting techniques	Restore degraded landscapes	
Extreme Weather Events	Increased frequency and intensity of storms, wildfires, and windthrow	Create buffer zones and firebreaks	Promote controlled burning and wildfire management practices	Schuldt et al. (2020)
	Direct damage to forest structure and biodiversity	Implement early warning systems and rapid response strategies	Encourage mixed-species plantations to increase resilience	
Elevated CO_2 concentrations	Altered photosynthesis rates and forest productivity	Monitor ecosystem changes and adjust management plans accordingly	Support research and development in carbon capture technologies	Lykhovyd (2021)
	Potential shifts in species competition and ecosystem dynamics	Promote mixed-species forests to enhance resilience	Implement sustainable harvesting practices	
Sea level rise	Salinization of coastal forests and loss of mangroves	Protect and restore coastal ecosystems	Implement sustainable coastal zone management	Anu et al. (2024)
	Increased erosion and loss of habitat	Utilize mangrove restoration and afforestation	Develop green infrastructure in coastal areas	
Changes in biodiversity	Loss of species and genetic diversity	Establish protected areas and conservation corridors	Promote ex-situ conservation efforts like seed banks	Lukina et al. (2021)
	Disruption of ecosystem services (e.g., pollination, water purification)	Engage in community-based forest management	Implement biodiversity-friendly land use practices	
Socio-economic impacts	Reduced availability of forest resources for local communities	Develop alternative livelihood programs	Promote community-based forest monitoring	Prochazka et al. (2023)
	Increased conflict over land and resource use	Foster participatory approaches in forest governance	Support sustainable tourism and eco-friendly enterprises	

(ii) Reforestation and Afforestation: Promoting reforestation and afforestation with climate-resilient species and genotypes to restore degraded areas and expand forest cover.

(iii) Fire Management: Developing and implementing effective fire management strategies, including controlled burns and firebreaks, to reduce the risk and impact of wildfires.

In addition to these strategies, integrating adaptive forest management practices that consider future climate scenarios is crucial for long-term forest resilience. This can include enhancing forest connectivity through the creation of ecological corridors, which facilitate species migration and genetic exchange in response to shifting climatic conditions (Heller & Zavaleta, 2009). Moreover, incorporating traditional ecological knowledge from indigenous and local communities can offer valuable insights into sustainable forest management practices that have been refined over generations and are often well-suited to local environmental conditions (Lake et al., 2017). The use of monitoring and modeling tools to predict climate impacts on forests enables managers to make informed decisions and implement proactive measures, such as adjusting thinning regimes or altering species compositions in plantations, to foster forest ecosystems that are better equipped to withstand climate-related stressors (Lindner et al., 2010). By combining these approaches, forest management can not only mitigate the adverse effects of climate change but also enhance the ecological, economic, and social benefits that forests provide.

4.2 Conservation Efforts

Conservation strategies to protect forest ecosystems from climate change impacts include expanding and effectively managing protected areas to conserve critical habitats and biodiversity (Börner et al., 2020). Additionally, establishing ecological corridors can facilitate species migration and adaptation to changing climates by enhancing connectivity within fragmented landscapes (Gregory et al., 2021). Moreover, monitoring and managing invasive species is crucial to prevent their spread and mitigate their impact on native ecosystems, ensuring the resilience of forest ecosystems in the face of climate change (Panzavolta et al., 2021).

4.3 Policy and Governance

Effective policy and governance are essential for addressing climate change impacts on forests:

(i) Climate Policies: Integrating forest conservation and management into national and international climate policies, such as the Paris Agreement, to ensure forests are part of climate mitigation and adaptation strategies (Moomaw et al., 2020).

(ii) Community Involvement: Engaging local communities and indigenous peoples in forest management and conservation efforts to leverage traditional knowledge and ensure sustainable practices (Dawson et al., 2021).

(iii) Financial Mechanisms: Developing financial mechanisms, such as carbon credits and payment for ecosystem services (PES), to incentivize forest conservation and sustainable management (Karsenty et al., 2014).

(iv) Research and Monitoring: Continued research and monitoring are crucial for understanding and addressing the impacts of climate change on forests, which may include conducting long-term ecological studies to track changes in forest ecosystems and identify trends and emerging threats (Kauppi et al., 2022). Utilizing climate models to predict future impacts on forests and guide adaptive management strategies is also the need of the hour. It is also imperative to utilize technological advancements like remote sensing and geographic information systems (GIS) to monitor forest health and detect changes in real time.

5 Conclusion

The impacts of climate change on forest cover are profound and multifaceted, encompassing ecological, socio-economic, and cultural dimensions. Addressing these challenges requires a comprehensive approach that integrates sustainable management practices, conservation efforts, effective policy and governance, and ongoing research and monitoring. By adopting these strategies, we can enhance the resilience of forest ecosystems, protect biodiversity, and ensure that forests continue to provide their invaluable services in the face of a changing climate. As we move forward, it is imperative that we recognize the intrinsic value of forests and the critical role they play in our global ecosystem. Through concerted efforts at local, national, and international levels, we can mitigate the adverse effects of climate change on forest cover and foster a sustainable future for both forests and the communities that depend on them.

References

Adak, S., Mandal, N., Mukhopadhyay, A., Maity, P. P., & Sen, S. (2023). Current state and prediction of future global climate change and variability in terms of CO_2 levels and temperature. In *Enhancing resilience of dryland agriculture under changing climate: Interdisciplinary and convergence approaches* (pp. 15–43). Springer Nature Singapore.

Albrich, K., Rammer, W., & Seidl, R. (2020). Climate change causes critical transitions and irreversible alterations of mountain forests. *Global Change Biology, 26*(7), 4013–4027. https://doi.org/10.1111/gcb.15118

Anu, K., Sneha, V. K., Busheera, P., Muhammed, J., & Augustine, A. (2024). Mangroves in environmental engineering: Harnessing the multifunctional potential of Nature's coastal architects for sustainable ecosystem management. *Results in Engineering*, 101765.

Artaxo, P., Hansson, H. C., Machado, L. A. T., & Rizzo, L. V. (2022). Tropical forests are crucial in regulating the climate on Earth. *PLOS Climate, 1*(8), Article e0000054. https://doi.org/10.1371/journal.pclm.0000054

Balthazar, V., Vanacker, V., Molina, A., & Lambin, E. F. (2015). Impacts of forest cover change on ecosystem services in high Andean mountains. *Ecological Indicators, 48*, 63–75. https://doi.org/10.1016/j.ecolind.2014.07.043

Baron, V. S., & Bélanger, G. (2020). Climate, climate-change and forage adaptation. *Forages: The Science of Grassland Agriculture, 2*, 151–186.

Bonan, G. B. (2008). Forests and climate change: Forcings, feedbacks, and the climate benefits of forests. *Science, 320*(5882), 1444–1449. https://doi.org/10.1126/science.1155121

Börner, J., Schulz, D., Wunder, S., & Pfaff, A. (2020). The effectiveness of forest conservation policies and programs. *Annual Review of Resource Economics, 12*, 45–64. https://doi.org/10.1146/annurev-resource-110119-025703

Carlson, C. J., Albery, G. F., Merow, C., Trisos, C. H., Zipfel, C. M., Eskew, E. A., Olival, K. J., Ross, N., & Bansal, S. (2022). Climate change increases cross-species viral transmission risk. *Nature, 607*(7919), 555–562. https://doi.org/10.1038/s41586-022-04788-w

Dawson, N.M., Coolsaet, B., Sterling, E.J., Loveridge, R., Gross-Camp, N.D., Wongbusarakum, S., Sangha, K.K., Scherl, L.M., Phuong Phan, H., Zafra-Calvo, N., & Lavey, W.G. (2021). The role of indigenous peoples and local communities in effective and equitable conservation. https://doi.org/10.5751/ES-12625-260319

De Frenne, P., Lenoir, J., Luoto, M., Scheffers, B. R., Zellweger, F., Aalto, J., et al. (2021). Forest microclimates and climate change: Importance, drivers and future research agenda. *Global Change Biology, 27*(11), 2279–2297. https://doi.org/10.1111/gcb.15569

Dotaniya, M. L. (2015). Impact of rising atmospheric CO_2 concentration on plant and soil process. In *Crop growth simulation modelling and climate change* (pp. 69–86). Scientific Publisher.

Ela, M. Z., Rabya, T., Khan, L., Rahman, M. H., Shovo, T. E. A., Jahan, N., Hossain, M. T., & Islam, M.N. (2021). Climate change and livelihood vulnerabilities: The forest resource-dependent communities of the sundarbans of Bangladesh. In *Climate vulnerability and resilience in the global south: human adaptations for sustainable futures* (pp. 341–352).

FAO. (2020). State of the world's forests 2020: Forests, biodiversity and people. Food and Agriculture Organization of the United Nations. http://www.fao.org

Felton, A. J., Knapp, A. K., & Smith, M. D. (2021). Precipitation–productivity relationships and the duration of precipitation anomalies: An underappreciated dimension of climate change. *Global Change Biology, 27*(6), 1127–1140. https://doi.org/10.1111/gcb.15480

Gregory, A., Spence, E., Beier, P., & Garding, E. (2021). Toward best management practices for ecological corridors. *Land, 10*(2), 140. https://doi.org/10.3390/land10020140

Hartmann, H., Bastos, A., Das, A. J., Esquivel-Muelbert, A., Hammond, W. M., Martínez-Vilalta, J., et al. (2022). Climate change risks to global forest health: Emergence of unexpected events of elevated tree mortality worldwide. *Annual Review of Plant Biology, 73*, 673–702. https://doi.org/10.1146/annurev-arplant-102820-012804

Hasan, S. S., Zhen, L., Miah, M. G., Ahamed, T., & Samie, A. (2020). Impact of land use change on ecosystem services: A review. *Environmental Development, 34*, Article 100527. https://doi.org/10.1016/j.envdev.2020.100527

Heller, N. E., & Zavaleta, E. S. (2009). Biodiversity management in the face of climate change: A review of 22 years of recommendations. *Biological Conservation, 142*(1), 14–32. https://doi.org/10.1016/j.biocon.2008.10.006

Hergarten, M., Liagre, L., & Froede-Thierfelder, B. (2013). *Forests and climate change adaptation: a twofold approach*. ECO Consult.

Hurteau, M. D. (2021). The role of forests in the carbon cycle and in climate change. In *Climate change* (pp. 561–579). Elsevier. https://doi.org/10.1016/B978-0-12-821575-3.00027-X

IPCC. (2018.) Annex I: Glossary [Matthews JBR (ed)]. In V. Masson-Delmotte, et al (Eds.), *Global warming of 1.5°C. An IPCC special report on the impacts of global warming of 1.5°C above pre-industrial levels and related global greenhouse gas emission pathways, in the context of strengthening the global response to the threat of climate change, sustainable development, and efforts to eradicate poverty* (pp. 541–562). Cambridge University Press. https://doi.org/10.1017/9781009157940.008

Karsenty, A., Vogel, A., & Castell, F. (2014). "Carbon rights", REDD+ and payments for environmental services. *Environmental Science & Policy, 35*, 20–29. https://doi.org/10.1016/j.envsci.2012.08.013

Kauppi, P. E., Stål, G., Arnesson-Ceder, L., Sramek, I. H., Hoen, H. F., Svensson, A., et al. (2022). Managing existing forests can mitigate climate change. *Forest Ecology and Management, 513*, Article 120186. https://doi.org/10.1016/j.foreco.2022.120186

Lake, F. K., Parrotta, J. A., & Giardina, C. P. (2017). Integration of traditional and western knowledge in forest landscape restoration. In *Forest landscape restoration: Integrating natural and social sciences* (pp. 111–140). Springer. https://doi.org/10.1007/978-3-319-69593-0_6

Lawrence, D., Coe, M., Walker, W., Verchot, L., & Vandecar, K. (2022). The unseen effects of deforestation: Biophysical effects on climate. *Frontiers in Forests and Global Change, 5*, 49. https://doi.org/10.3389/ffgc.2022.756115

Leon, M., Cornejo, G., Calderón, M., González-Carrión, E., & Florez, H. (2022). Effect of deforestation on climate change: A co-integration and causality approach with time series. *Sustainability, 14*(18), 11303. https://doi.org/10.3390/su141811303

Lindner, M., Maroschek, M., Netherer, S., Kremer, A., Barbati, A., et al. (2010). Climate change impacts, adaptive capacity, and vulnerability of European forest ecosystems. *Forest Ecology and Management, 259*(4), 698–709. https://doi.org/10.1016/j.foreco.2009.09.023

Lukina, N. V., Geraskina, A. P., Gornov, A. V., Shevchenko, N. E., Kuprin, A. V., Chernov, T. I. et al. (2021). Biodiversity and climate-regulating functions of forests: current issues and research prospects. *Вопросы лесной науки, 4*(1), 1. https://doi.org/10.31509/2658-607x-202141k-60

Lykhovyd, P. V. (2021). Role of forests in regulation of climate and crops productivity. *Plant Archives, 21*(1), 731–735.

McDowell, N. G., Allen, C. D., Anderson-Teixeira, K., Aukema, B. H., Bond-Lamberty, B., Chini, L., Xu, C. (2020). Pervasive shifts in forest dynamics in a changing world. *Science, 368*(6494), eaaz9463. https://doi.org/10.1126/science.aaz9463

McLaughlin, D. L., Kaplan, D. A., & Cohen, M. J. (2020). Managing forests for increased water yield in the southeastern U.S. coastal plain. *Journal of the American Water Resources Association, 56*(1), 101–116. https://doi.org/10.1111/1752-1688.12806

Moomaw, W. R., Law, B. E., & Goetz, S. J. (2020). Focus on the role of forests and soils in meeting climate change mitigation goals: Summary. *Environmental Research Letters, 15*(4), Article 045009. https://doi.org/10.1088/1748-9326/ab6b38

Morgan, E. A., Cadman, T., & Mackey, B. (2021). Integrating forest management across the landscape: A three pillar framework. *Journal of Environmental Planning and Management, 64*(10), 1735–1769. https://doi.org/10.1080/09640568.2020.1837747

Nias, R. C., & Mooney, J. R. (2013). Endangered ecosystems. *Encyclopaedia of Biodiversity, 2*, 407–424. https://doi.org/10.1016/B978-0-12-384719-5.00257-4

Ontl, T. A., Janowiak, M. K., Swanston, C. W., Daley, J., Handler, S., Cornett, M., et al. (2020). Forest management for carbon sequestration and climate adaptation. *Journal of Forestry, 118*(1), 86–101. https://doi.org/10.1093/jofore/fvz062

Panzavolta, T., Bracalini, M., Benigno, A., & Moricca, S. (2021). Alien invasive pathogens and pests harming trees, forests, and plantations: Pathways, global consequences and management. *Forests, 12*(10), 1364. https://doi.org/10.3390/f12101364

Prochazka, P., Abrham, J., Cerveny, J., Kobera, L., Sanova, P., Benes, D., et al. (2023). Understanding the socio-economic causes of deforestation: A global perspective. *Frontiers in Forests and Global Change, 6*, 1288365. https://doi.org/10.3389/ffgc.2023.1288365

Schuldt, B., Buras, A., Arend, M., Vitasse, Y., Beierkuhnlein, C., Damm, A., et al. (2020). A first assessment of the impact of the extreme 2018 summer drought on Central European forests. *Basic and Applied Ecology, 45*, 86–103. https://doi.org/10.1016/j.baae.2020.04.003

Sesana, E., Gagnon, A. S., Ciantelli, C., Cassar, J., & Hughes, J. J. (2021). Climate change impacts on cultural heritage: A literature review. *Wiley Interdisciplinary Reviews: Climate Change, 12*(4), Article e710. https://doi.org/10.1002/wcc.710

Uriarte, M., Thompson, J., & Zimmerman, J. K. (2019). Hurricane María tripled stem breaks and doubled tree mortality relative to other major storms. *Nature Communications, 10*, 1362. https://doi.org/10.1038/s41467-019-09319-2

Weiskopf, S. R., Rubenstein, M. A., Crozier, L. G., Gaichas, S., Griffis, R., Halofsky, J. E., et al. (2020). Climate change effects on biodiversity, ecosystems, ecosystem services, and natural resource management in the United States. *Science of the Total Environment, 733*, Article 137782. https://doi.org/10.1016/j.scitotenv.2020.137782

Xiao, J. L., Zeng, F., He, Q. L., Yao, Y. X., Han, X., & Shi, W. Y. (2021). Responses of forest carbon cycle to drought and elevated CO_2. *Atmosphere, 12*(2), 212.

Impact of Land Use Change on Disease Outbreak

Sanchari Burman, Sushmita Suman Sinha, LVSairam K. Nagalla, and Shweta Meshram

Abstract

The impact of land use on disease outbreaks is a critical area of study that underscores the intricate relationship between human activities and ecological balance. Intentional modifications of biophysical characteristics, such as deforestation, urbanization, and agricultural conversion, significantly influence disease dynamics. These changes create new habitats for vectors, alter wildlife behavior, and increase the transmission of zoonotic and vector-borne diseases. Historical narratives and modern frameworks like the Land-Use and Land-Cover Change (LUCC) Core Project highlight the urgency of addressing unsustainable practices. Socioeconomic factors and human behavior further complicate these dynamics, making it essential to integrate public health strategies with sustainable land management. This chapter emphasizes the need for interdisciplinary research to develop effective mitigation strategies and policies to reduce the health risks associated with land use changes.

Keywords

Land use change · Disease outbreaks · Deforestation · Urbanization · Socioeconomic factors · Sustainable land management

S. Burman · S. S. Sinha
Institute of Agricultural Sciences, Banaras Hindu University, Varanasi, India

L. K. Nagalla · S. Meshram (✉)
School of Agriculture, Lovely Professional University, Punjab, India
e-mail: shweta.26662@lpu.co.in

1 Introduction

The FAO defines land as the surface of the earth with the biosphere that surrounds it, including the climate, soil, hydrology, sedimentary layers, populations of plants and animals, and human settlements and the activities that go along with them. According to Briassoulis (2009), land use is the intentional modification of biophysical characteristics. According to the FAO, land use also refers to how human activity directly affects land resources. It could relate to a single land cover (pastoralism, for example), support several uses (forest for agriculture, logging, and recreation, for example), or include maintaining different land covers within a single system (farming, which combines cultivated land, woodlots, pasture, and settlements). Examples of land-use conversion include alterations to the distribution and pattern of land uses in a given area.

Land-use changes can also include alterations to a particular form of land use, such as changes to its intensity or distinguishing characteristics.

To accommodate a variety of requirements human civilization has changed the environment significantly over time. This primarily human-caused change has been more intense since the 1700s and, has hurt society and the environment. Changes in land use concern human goals and impact the environment. The changing of the ecosystem into human-required farmland, industrial areas, and urban areas, mining and deforestation affect ecological balance and result in land degradation. As the biodiversity and ecosystem have been changed the transmission of diseases. Potential habitat suitability, and recreation of new disease reservoirs with new ecological niches due to this land use change make the relation between land use change and disease outbreak more widely recognized.

Historical narratives, such as influential books like Marsh's "Man and Nature," have emphasized the significant effects

J. A. Parray (ed.), *Soil and Land Use Change*, Sustainable Landscape Planning and Natural Resources Management,
https://doi.org/10.1007/978-3-031-93075-1_12

of human activity on ecosystems and landscapes. Concepts such as the Land-Use and Land-Cover Change (LUCC) Core Project have been popular in the modern era due to the pressing need to address unsustainable land use practices. Environmental conservation and socio-economic development are crucial to managing land resources to decrease outbreaks and global health security. To reduce infectious hazards, we must emphasize the complex interrelation between the LUC and Disease Dynamics in an increasingly linked world, policies, interventions, and public health strategies. Through the comprehensive strategy implementation, communities can work towards a future that is more robust and sustainable (Briassoulis, 2009).

Globally, economic development and population expansion have led to notable changes in land use. The last few decades have seen a more significant and pervasive impact although there have been occasional shifts in land usage over the previous few years resulting in global warming and biodiversity loss. Intentional modifications in land use have led to habitat loss, decreased water retention, and soil erosion. Deforestation accounts for about 25% of the enhanced greenhouse effect caused by changes in land use. The assessment places particular emphasis on the widespread modification of forests, particularly tropical forests, which experience the highest rates of land-use change. The Global Range of Land Use Change The effects of land-use change have spread throughout the world during the last several centuries, especially in the last few decades (Houghton, 1994).

Deforestation, urbanization, and agriculture are examples of land use changes that have a substantial impact on the transmission of zoonotic illnesses, since 1940, land conversion has been connected to about half of all new cases of zoonotic diseases. The theories of disease transmission support a decline in biodiversity while the diverse hosts may magnify emerging infections, the dilution effect leads to the spread of illness. The difference is based on the infections, fauna, and environment that are involved (Bjornstad, 2021).

Many environmental conditions can affect the incidence and spread of vector-borne diseases (VBDs) by directly affecting vector behavior or pathogen transmission. Sala et al. (2000) predicted that land-use intensification would impact biodiversity loss; however, several reviews and correlation studies have found less evidence to bolster this assertion. We looked at the features of climate change that might affect the risk of VBD transmission to fill this information gap, being especially mindful of the distinction between LULC effects and climate change. Since how land-use intensity is mediated can affect the impact of pathogens on transmission and infectivity, the final target of the pathogens, i.e., hosts versus vectors, were studied separately (Ferraguti et al., 2023).

2 Classifying Land Use Change

Our globe is being divided in large part and significantly impacts biodiversity, the environment, and human well-being. To reduce any negative effects and ensure sustainable management, it is mandatory to categorize these changes. The three main categories of LUC—deforestation, urbanization, and cultivation—have been the subject of recent research (Bjornstad, 2021).

2.1 Conversion: Changing of Land Use Form

Conversion means the fundamental transformation of land from one land use category to another (Liu et al., 2022). Converting a forest into cultivable land or rural areas into cities.

2.1.1 Deforestation

Cutting forests for industry, agriculture, or settlements. Deforestation raises the edge density, which leads to increased animal-human links and an increase in the likelihood of zootonic disease transmission. Research conducted in Madagascar and Uganda shows how habitat degradation modifies animal behavior and promotes the spread of illness by habitat degradation modifying animal behavior. Deforestation also increases mosquito populations, which benefits disease vectors. This establishes the connection between disease emergence and land conversion (Bjornstad, 2021).

2.1.2 Urbanization

Transformation of rural land to accommodate urban growth. However, vector populations can create new breeding grounds in highly developed areas. Also, the lower level of urbanization can cause increased disease transmission (Bjornstad, 2021). A developed country with a well-organized healthcare system can reduce disease outbreaks but at the same time, a country that has a less developed health system can increase disease transmission (Kalbus et al., 2021). Poor drainage systems in urban areas may cause floods, and flooding degrades the water quality that becomes a bed for pathogens. Land use changes due to urbanization are ever-expanding as the population increases and spreads disease outbreaks (Table 1).

2.1.3 Cultivation

Agriculture can spread disease because it establishes new vector breeding grounds. Terraforming the terrain or establishing irrigation can result in the formation of new breeding grounds. The rising population of cattle creates new feeding reservoirs. Additionally, by changing their behaviors and compositions, the surrounding fauna may be impacted by

Table 1 Impact of urbanization on disease outbreak

Examples	References
In Malawi, small and marginal-level agriculture in more developed areas expands the abundance of vectors by generating new microclimates for vector development	Bjornstad (2021)
Due to the change in microclimate, transmission of diseases like malaria has increased in the Amazon region as well as in Ecuador	Bjornstad (2021)
Due to unintended urbanization in Central and South America new larval habitats have been created. The newly established habitat where there is no predator, helps to expand the dengue mosquito population as they were able to breed in water containers, drains, and gutters	Kalbus et al. (2021)
It is additionally more possible for people living in rural areas to come into contact with diseased wildlife during their regular activities. Raising chickens and keeping mice or shrews in the home were the most common contact points that resulted in seropositivity of bat-borne illnesses 15 across rural Southern China	Li et al. (2019)

Table 2 Describing the impact of cultivation on disease outbreak

Examples	Reference
The presence of clean water in Asia boosted the population of the lymphatic filariasis vector, while the cleared land created more stagnant puddles, which in turn boosted the population of the Japanese encephalitis vector	Franklinos et al. (2019)
Papyrus wetlands in Uganda's southwest highlands were drained and put to use for cultivation in the early 1940s. There was a malaria outbreak in 1946 not long after this. Because of the lower temperatures, which make mosquito larvae unfriendly, highland areas are generally immune to malaria outbreaks. Because the highland population lacks an immune to the disease, they are therefore more susceptible to a malarial outbreak than are lowland populations	Zsuffa et al. (2014)
The raising of livestock expands the amount of blood that mosquitoes have access to, creating a potentially effective reservoir that may sustain or magnify zoonotic infections. In Asia, pig rearing increased the spread of a zoonotic disease that causes Japanese encephalitis. Because the pigs were able to develop large virus concentrations, there were more infected vectors in the vicinity	Burkett-Cadena and Vittor (2018); Franklinos et al. (2019)
Over the past few decades, land agriculture in East Africa has increased by more than 70%. A mouse species known as *Mastomysnatalensis*, which is extremely adept at handling the plague Y. pestis, was discovered to be 20 times more prevalent in agricultural locations in Northern Tanzania than in core habitats	McCauley et al. (2015)

cultivation. Irrigation and intentionally constructed dams to facilitate crops might generate new areas for vector reproduction (Bjornstad, 2021) (Table 2).

3 Vector Dynamics and Land Use

Vector-borne diseases (VBDs) are infectious diseases spread by blood-feeding arthropods (the vectors), such as fleas, ticks, and mosquitoes, to humans or wildlife hosts. They are widely acknowledged to be a serious global public health issue with substantial financial ramifications (Kilpatrick & Randolph, 2012; Franklinod et al., 2019).

A bite from an infected hematophagous arthropod, such as a mosquito, tick, sandfly, midge, triatomine bug, or blackfly, can spread parasites, germs, or viruses to humans and other vertebrates. Approximately 6.3 billion people are at risk for vector-borne infections, which cause more than 300 million instances of illness and over 700,000 deaths annually. Dengue fever and malaria currently have the highest rates of disease. Plasmodium spp. parasites, which are spread by *Anopheles* mosquitoes, are the cause of malaria. Every year, there are 219 million cases worldwide and over 500,000 fatalities, with children in Africa being disproportionately affected. The dengue virus (DENV), a significant arbovirus that infects humans, is spread by *Aedes* (*Stegomyia*) mosquitoes, primarily *Aedes aegypti* but also occasionally *Aedes albopictus*. They represent 96 million cases (Al Nazawi et al., 2019).

3.1 Vector Dynamics Impact on Plants

Although the diseases are caused by viruses (symptoms appear like the afflicted plant is yellow and discolored), the transmissible vector is an alate aphid. Chemical cues and volatile organic compounds that vectors can detect and the changes inspectral reflectance due to the change in symptoms. These are which help the vector to identify and move toward the infected plant for transmission to take place. The vector must engage in tropically interacting with the infected host (probing, biting, or prolonged feeding) to obtain the infection. The physiological, and biochemical characteristics of the host and the changes in the defense brought by the infection can have an impact on the type of trophic interaction (Lin et al., 2019).

3.2 Plant Pathogenic Nematodes

Tobacco rattle Virus which is an economically significant disease of farmed potatoes (*Solanum tuberosum*) declines crop quality and renders the crop unmarketable (Sahi et al., 2015). Displaying a *Trichodorus* worm, which is the TRV vector, and necrotic rings, which are typical of spraing disease/corky ringspot are the effects of TRV infection.

3.3 Human Ability and Behavior

Due to national and international movements, the spread of vectors and associated infections into areas with favorable environmental conditions become very frequent (Kraemer et al., 2015). For instance, it is likely that throughout the sixteenth and seventeenth centuries, the slave trade brought YFV and *A. aegypti* from Africa to the Americas (Chippaux and Chippaux, 2018; Cuthbert et al., 2023).

The introduction of ticks is frequently related to vertebrate hosts used in agriculture, such as *Amblyomma variegatum*, *Haemaphysalis longicornis*, and *Rhipicephalus microplus*. These species were probably introduced into the Americas during the eighteenth and nineteenth centuries (Cuthbert et al., 2023). Numerous vector-borne diseases implicate households as the primary sites of infection. It has long been believed that one of the main risk factors for illnesses spread by mosquitoes, such as DENV and CHIKV, is poverty. Unfortunately, few studies consistently quantify social vulnerability and income, thus this association is debatable (Mohd Royali, 2023). By increasing the financing for interdisciplinary research regarding the mechanisms of transmission of vector-borne diseases concerning the climate.

3.4 Land Use and Land Cover Changes

Agriculture, exploitation of natural resources, industries and man-made construction on the surface of land are the main reasons for land degradation that define land use change. These actions have the potential to greatly exacerbate climate change by reducing natural carbon capture and storage and causing a loss of biodiversity. Furthermore, it has been observed that changes in anthropogenic land use and cover are a major factor in the transmission of zoonotic infections from wildlife to human populations. (Song et al., 2018; Gibb et al., 2020).

Changes in land use and cover can impact vector-borne diseases by influencing vector-host contact rates, microclimate suitability for vectors and pathogens, vector predators, vector larval and adult habitats, or a mix of these variables. Circular pairs of upward and downward arrows indicate the likelihood of an increase or decrease brought on by anthropogenic factors such as mining, dam construction, agriculture, animal production, urbanization, and deforestation. Vertical arrows imply that a more gradual shift is anticipated.

4 Human Behavior and Land Use Patterns

Examine how human conduct impacts the transmission of illness in diverse land use contexts. Consider the impact of socioeconomic status on health outcomes.

Animal relationships and habitat patterns are altered as humans alter landscapes, whether through extensive conversion of natural habitats or more focused and localized activities such artisanal gold mining, hunting, and selective logging. According to Foley et al. (2005) and Glidden et al. (2021), all of these changes may affect the transmission of vector-borne diseases (VBDs).

As vectors and infections often occupy distinct niches, disease systems can respond differently to environmental changes during each cycle of transmission. It is reasonable to expect changes in the incidence of specific diseases with a rise in human activity, for instance, disease systems that are well-suited to mosaics of agriculture could give way to disease systems that thrive in urban areas. However, our incomplete understanding of the connections between different VBDs and human–environment interactions hinders projections of how disease assemblages collectively fluctuate across complex and changing contexts (Kilpatrick & Randolph, 2012). One technique to look at land-use and land-cover classes is to use broad classifications like "urban area" or "forest area." The effects of land-use change on VBD transmission (Bjornstad, 2021).

4.1 Socioeconomic Factors

According to Bhutta et al. (2014), socioeconomic variables influence people's lifestyle, quality of life, and interactions with the environment, which in turn affects how human diseases arise and persist. The realized ecological niche of parasite species can be found using the framework for modelling ecological niches. As proxies for the impacts of physiological tolerance, biological interactions, and dispersal on the ecology of animals, realized ecological niches are measured as a hypervolume along environmental variables inhabited by the species (e.g., pathogens, vectors) (Escobar et al., 2020).

4.2 Effect of Land Use Change

In many parts of the world, there are plenty of natural mosquito habitats—even in the absence of human environmental change. Regardless of the motivation, human alteration of the environment such as clearing land for subsistence farming or building dams for recreational and hydroelectric power often makes mosquito-related issues worse by generating new habitats, expanding existing ones, or changing them in a way that could cause limited mosquito populations to explode when new habitats become available. It has long been known that human-caused land-use changes contribute to the worsening of diseases spread by mosquitoes. While the impacts of other human-induced environmental alterations are only now being identified or rediscovered, many types of habitat disruption have been extensively researched and documented. These changes fall under several broad categories.

4.3 Agricultural Development

The rice paddy is one of the most significant agricultural changes linked to diseases spread by mosquitoes. Immature mosquito development thrives in these shallow pools of water with emergent plants and minimal movement. Contiguous rice paddies can cover enormous areas in various parts of the world, leading to unthinkably high mosquito concentrations. Large-scale rice fields are known to be associated with higher vector densities; however, the impacts on malaria transmission seem to rely on immunity within the surrounding population (Ljumba & Lindsay, 2001).

4.4 Public Health

The results show that, but not for *Rickettsia typhi*, exposure to agriculture is strongly correlated with either scrub typhus or the rickettsioses that cause spotted fever. This disparity might again have to do with transmission characteristics. whereas rickettsioses and scrub typhus are also vector-borne typhus-based disorders, spotted fever rickettsioses are carried by fleas, whereas scrub typhus is carried by ticks.

This is consistent with recent studies that indicate ticks are extremely sensitive to changes in their surroundings (Dantas-Torres, 2015; Beugnet & Chalvet-Monfray, 2013).

For instance, in North America, the prevalence of tick-borne illnesses like Lyme disease has grown due to forest fragmentation (Killilea et al., 2008).

5 Mitigation Strategies

According to Gai and Wang (2024), there is a growing risk to primary productivity, global food security, and biodiversity loss in many vulnerable locations of the world due to the increased outbreaks and severity of plant diseases. The consequences of these disease epidemics include losses to yield and the environment. For example, pathogens—microorganisms that cause diseases and impair host health and productivity—and pests alone are estimated to be responsible for US$220 billion of the annual loss in crop yield (Hossain et al., 2024). This has a direct impact on the surrounding socioeconomic issues, the local economy, and food security (Muluneh, 2021) Pathogenic microorganisms such as *Penicillium* spp. and *Xanthomonas euvesicatoria* cause post-harvest loss.

5.1 Technology Developments to Track, Control, and Lower the Risks of Illness in the Context of Global Development

Limits are single pathogen concentration, single crop and single disease paradigms for measuring reducing and managing disease risks. For comprehending pathogen biology and ecology, from the molecular to the global, interdisciplinary research is needed. Integrating biological data with variables like trade, transportation, geography, and climate might enhance surveillance and forecast disease. Improved early disease reporting through the use of artificial intelligence, biochemical sensors, satellite and remote sensing technology, permanent observatories that track airborne signals and vectors, and farmers' and other volunteers' participation can all contribute to effective surveillance (Garrett et al., 2022) (Fig. 1).

Plant disease management requires a coordinated effort combining technological advances in artificially generated microbes, precision agriculture and agroecological methods to create workable and long-lasting solutions. Drones, AI, and machine learning-based precision agriculture technologies can assist targeted management which is safe for the environment and helps in the early detection of diseases. Recent research in the fields of microbiology (microbiome engineering approaches), biochemistry (volatiles and plant elicitors), and synthetic biology (Synthetic Microbial Community, or SynCom) offer new approaches to managing diseases.

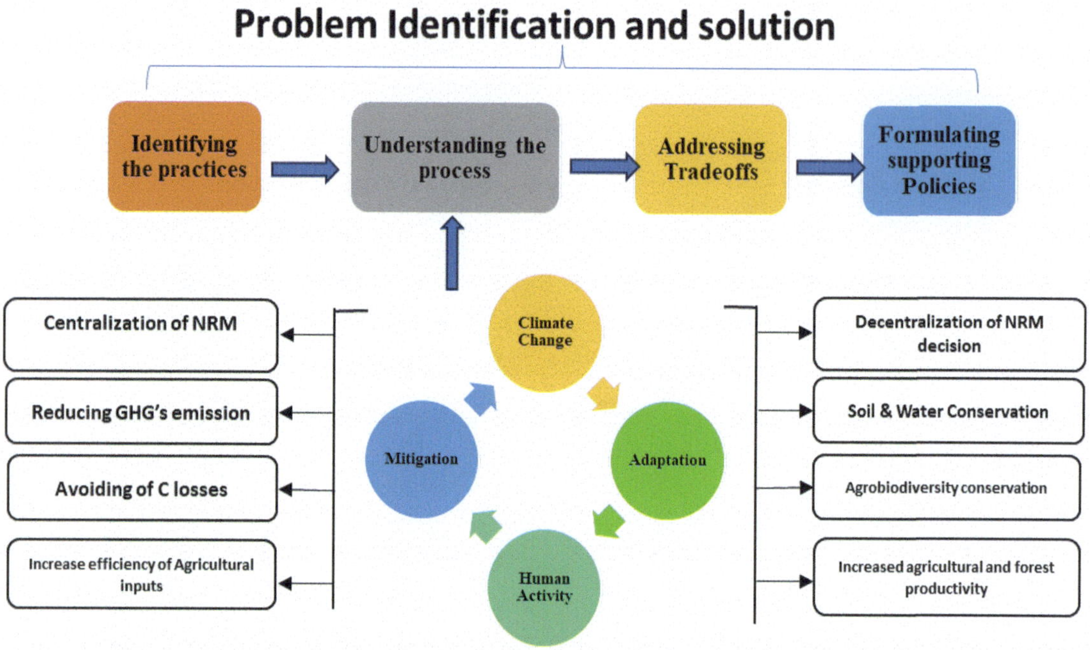

Fig. 1 Strategies of mitigation and risk analysis

5.2 Interface Between Science and Policy and Social Innovations

An effective system for tracking, regulating, and lessening the consequences of plant diseases is essential for maintaining the sociopolitical stability of a country or an area. One of the finest examples is the potato Phytophthora blight, which caused the Irish famine of 1859, which claimed two million lives and caused extensive migrations. This tragedy was a terrible outcome of the absence of efficient measures to control risks linked with illness. Other instances include the recent coffee rust outbreak that resulted in a more than 50% loss in coffee productivity and caused starvation, poverty, and a mass exodus from Central America, as well as the famine and deaths connected to brown spot disease in rice in Bengal, India (Ristaino et al., 2021).

5.3 Interfaces Among Science, Policy, and Society

To be effective, networks such as the global surveillance system would need to create science-policy-society interfaces (SPSIs) that operate at the local, regional, and global levels. If these SPSIs were in continuous partnership with other organizations that are actively engaged at the intersection of science, policy, and society, they would provide assessment, monitoring, forecasting, and recommendations for policy actions that take science, non-traditional knowledge (such as the indigenous and private sectors), and socioeconomic conditions into account.

Given their wide skill set and institutional locations in developing nations where new data are needed, established organizations like the Consultative Group for International Agricultural Research (One CGIAR) would be well-positioned to coordinate a larger knowledge network.

If given a sufficient mandate and resources, one CGIAR might coordinate periodic assessments, forecasting, and monitoring of plant diseases with the Food and Agriculture Organisation of the United Nations (FAO). This could be improved by utilizing the expertise of two organizations and the global research community: the Intergovernmental Panel on Climate Change (IPCC) and the International Platform for Biodiversity and Ecosystem Services (IPBES), which address climate change and biodiversity loss, respectively. Both sides will benefit from formalizing collaborations with these international organiz, such as One Health. This will make it possible for the SPSIs to estimate disease incidence linked to biodiversity loss, climate change, and the consequent effects on plant and human health using their models, data, and resources.

6 Future Scope in Soil Science and Agriculture- Disease Dynamics Relation

It has been demonstrated that land use changes significantly influence the spread of diseases. Several factors, including vegetation dynamics, climate change, and agricultural practices make a complex relationship, which affects land use changes and the dynamics of disease (Kumar et al., 2022; Wan et al., 2020).

Furthermore, Wan et al.'s (2020) research revealed that changes in land use significantly impact agriculture by influencing trends in vegetation distribution and climate changes. Studies have also shown that vegetation changes are significantly impacted by long-term interactions with numerous other changes in the environment and land use (Kumar et al., 2022). Cooperation between multiple academic fields is essential to fully understand how shifting land use influences disease outbreaks (Goldstein et al., 2022; Rulli et al., 2025). This cooperation will help determine how shifting land use affects the spread of diseases.

Researchers of different disciplines combine information to develop a comprehensive view of how land use, environmental shifts, and disease spread are co-related. To understand the critical pattern of disease transmission, we need to monitor the land use changes and establish protected areas. Studies show that careful land use management reduces the vector-transmitted disease outbreak (Eder et al., 2018; Kain et al., 2021). Additionally, the creation of new soil indices and the fusion of econometric theory with soil-environmental sciences highlight the transformative potential of interdisciplinary collaboration in advancing soil science research (Mizuta et al., 2018).

7 Conclusion

Future investigations into the effects of land use changes on disease outbreaks necessitate interdisciplinary collaboration to tackle the intricate challenges posed by evolving landscapes. Researchers can develop new ways to reduce the health hazards related to land use change through the compilation of different knowledge from various fields and different projects (Hou et al., 2020). The scientists of soil science and agriculture looking forward to better land use management, which will help to mitigate the consequences of land, use change on disease dynamics. So, for this, interdisciplinary research is necessary while encouraging projects also which play a very crucial role as it is a complex relationship and reveals the complex links between disease transmission, agricultural practices, and soil. Delving into sustainable agriculture and land use management is feasible to promote a better tomorrow (Hou et al., 2020).

References

Al Nazawi, A. M. (2019). *Behavioural and physiological insecticide resistance of the dengue vector Aedes aegypti in the Kingdom of Saudi Arabia*. The University of Liverpool (United Kingdom).

Beugnet, F., & Chalvet-Monfray, K. (2013). Impact of climate change in the epidemiology of vector-borne diseases in domestic carnivores. *Comparative Immunology, Microbiology and Infectious Diseases, 36*(6), 559–566.

Bhutta, Z. A., Sommerfeld, J., Lassi, Z. S., Salam, R. A., & Das, J. K. (2014). Global burden, distribution, and interventions for infectious diseases of poverty. *Infectious Diseases of Poverty, 3*, 1–7.

Bjornstad, A. (2021). *The relationship between land use change and disease prevalence*.

Briassoulis, H. (2009). Factors influencing land-use and land-cover change. *Land cover, land*

Burkett-Cadena, N. D., & Vittor, A. Y. (2018). Deforestation and vector-borne disease: Forest conversion favors important mosquito vectors of human pathogens. *Basic and Applied Ecology, 26*, 101–110.

Chippaux, J. P., & Chippaux, A. (2018). Yellow fever in Africa and the Americas: A historical and epidemiological perspective. *Journal of Venomous Animals and Toxins including Tropical Diseases, 24*, 20.

Cuthbert, R. N., Darriet, F., Chabrerie, O., Lenoir, J., Courchamp, F., Claeys, C., & Renault, D. (2023). Invasive hematophagous arthropods and associated diseases in a changing world. *Parasites & Vectors, 16*(1), 291.

Dantas-Torres, F. (2015). Climate change, biodiversity, ticks and tick-borne diseases: The butterfly effect. *International Journal for Parasitology: Parasites and Wildlife, 4*(3), 452–461.

Escobar, L. E. (2020). Ecological niche modeling: An introduction for veterinarians and epidemiologists. *Frontiers in Veterinary Science, 7*, 519059.

Ferraguti, M., Magallanes, S., Suarez-Rubio, M., et al. (2023). Does land-use and land cover affect vector-borne diseases? A Systematic Review and Meta-Analysis. *Landscecology, 38*, 2433–2451. https://doi.org/10.1007/s10980-023-01746-3

Foley, J. A., DeFries, R., Asner, G. P., Barford, C., Bonan, G., Carpenter, S. R., & Snyder, P. K. (2005). Global consequences of land use. *science, 309*(5734), 570–574.

Franklinos, L. H. V., Jones, K. E., Redding, D. W., & Abubakar, I. (2019). The effect of global change on mosquito-borne disease. *The Lancet Infectious Disease, 19*, 302–312.

Eder, M., Cortes, F., Teixeira de Siqueira Filha, N., Araújo de França, G. V., Degroote, S., Braga, C., ... & Turchi Martelli, C. M. (2018). Scoping review on vector-borne diseases in urban areas: transmission dynamics, vectorial capacity and co-infection. *Infectious Diseases of Poverty, 7*(1), 90.

Gai, Y., & Wang, H. (2024). Plant disease: A growing threat to global food security. *Agronomy, 14*(8), 1615.

Garrett, K. A., Bebber, D. P., Etherton, B. A., Gold, K. M., Plex Sulá, A. I., & Selvaraj, M. G. (2022). Climate change effects on pathogen emergence: Artificial intelligence to translate big data for mitigation. *Annual Review of Phytopathology, 60*, 357–378.

Gibb, R., Redding, D. W., Chin, K. Q., Donnelly, C. A., Blackburn, T. M., Newbold, T., & Jones, K. E. (2020). Zoonotic host diversity increases in human-dominated ecosystems. *Nature, 584*(7821), 398–402.

Glidden, C. K., Nova, N., Kain, M. P., Lagerstrom, K. M., Skinner, E. B., Mandle, L., & Mordecai, E. A. (2021). Human-mediated impacts on biodiversity and the consequences for zoonotic disease spillover. *Current Biology, 31*(19), R1342–R1361.

Goldstein, J. E., Budiman, I., Canny, A., & Dwipartidrisa, D. (2022). Pandemics and the human-wildlife interface in Asia: Land use change as a driver of zoonotic viral outbreaks. *Environmental Research Letters, 17*(6), 063009.

Hou, D., Bolan, N. S., Tsang, D. C. W., Kirkham, M. B., & O'Connor, D. (2020). Sustainable soil use and management: An interdisciplinary and systematic approach. *Science of the Total Environment, 729*, 138961.

Houghton, R. A. (1994). The worldwide extent of land-use change. *BioScience, 44*(5), 305–313.

Kain, M. P., MacDonald, A. J., Mordecai, E. A., & Mandle, L. (2021). *Land-use planning for health: Tradeoffs and nonlinearities govern how land-use change impacts vector-borne disease risk*. bioRxiv, 2021-06.

Kalbus, A., Sampaio, V. S., Boenecke, J., & Reintjes, R. (2013). Exploring the influence of deforestation on dengue fever incidence in the Brazilian Amazonas state. *PLOS ONE, 16*, 559–566.

Killilea, M. E., Swei, A., Lane, R. S., Briggs, C. J., & Ostfeld, R. S. (2008). Spatial dynamics of Lyme disease: A review. *EcoHealth, 5*, 167–195.

Kilpatrick, A. M., & Randolph, S. E. (2012). Drivers, dynamics, and control of emerging vector-borne zoonotic diseases. *The Lancet, 380*(9857), 1946–1955.

Kraemer, M. U., Sinka, M. E., Duda, K. A., Mylne, A. Q., Shearer, F. M., Barker, C. M., & Hay, S. I. (2015). The global distribution of the arbovirus vectors Aedes aegypti and Ae. albopictus. *elife, 4*, e08347.

Kumar, L., Chhogyel, N., Gopalakrishnan, T., Hasan, M. K., Jayasinghe, S. L., Kariyawasam, C. S., ... & Ratnayake, S. (2022). Climate change and future of agri-food production. In: *Future foods* (pp. 49–79). Academic Press.

Li, X., Liu, S., Cheng, Q., Wu, C., & Gong, P. (2019). Characterizing the dynamics of urban expansion in response to policy interventions: A case study of Guangzhou, China. *Landscape and Urban Planning, 189*, 149–162.

Lin, D., Xu, Y., Wu, H., Liu, X., Zhang, L., Wang, J., & Rao, Q. (2019). Plant defense responses induced by two herbivores and consequences for whitefly Bemisia tabaci. *Frontiers in Physiology, 10*, 346.

Liu, Y., Li, S., Fang, F., Zhao, Y., & Wang, S. (2022). A review on the classification and driving mechanisms of land-use changes. *Journal of Environmental Management, 308*, Article 114597.

Ljumba, J. N., & Lindsay, S. W. (2001). Impact of irrigation on malaria in Africa: Paddies paradox. *Medical and Veterinary Entomology, 15*, 1–11.

McCauley, D. J., Salkeld, D. J., Young, H. S., Makundi, R., et al. (2015). Effects of land use on plague (Yersinia pestis) activity in rodents in Tanzania. *The American Society of Tropical Medicine and Hygiene., 92*, 776–783.

Mohd Royali, M. S. (2023). *Community participation and risk communication for dengue prevention and control in Malaysia*.

Muluneh, M. G. (2021). Impact of climate change on biodiversity and food security: A global perspective—a review article. *Agriculture & Food Security, 10*(1), 1–25.

Nigusie, A., Gizaw, Z., Gebrehiwot, M., & Destaw, B. (2021). Vector-borne diseases and associated factors in the rural communities of northwest Ethiopia: A community-based cross-sectional study. *Environmental Health Insights, 15*, 11786302211043049.

Ristaino, J. B., Anderson, P. K., Bebber, D. P., Brauman, K. A., Cunniffe, N. J., Fedoroff, N. V., & Wei, Q. (2021). The persistent threat of emerging plant disease pandemics to global food security. *Proceedings of the National Academy of Sciences, 118*(23), Article e2022239118.

Rulli, M. C., D'Odorico, P., Galli, N., John, R. S., Muylaert, R. L., Santini, M., & Hayman, D. T. (2025). Land use change and infectious disease emergence. *Reviews of Geophysics, 63*(2), e2022RG000785.

Sahi, S. T., Imtia, M., & Dababat, A. A. (2015). The status of cereal nematodes in Pakistan. *Nematodes of Small Grain Cereals, 45*.

Sala, O. E., Stuart Chapin, F. I. I. I., Armesto, J. J., Berlow, E., Bloomfield, J., Dirzo, R., ... & Wall, D. H. (2000). Global biodiversity scenarios for the year 2100. *Science, 287*(5459), 1770–1774.

Song, X. P., Hansen, M. C., Stehman, S. V., Potapov, P. V., Tyukavina, A., Vermote, E. F., & Townshend, J. R. (2018). Global land change from 1982 to 2016. *Nature, 560*(7720), 639–643.

Zsuffa, I., Van Dam, A. A., Kaggwa, R. C., Namaalwa, S., Mahieu, M., Cools, J., & Johnston, R. (2014). Towards decision support-based integrated management planning of papyrus wetlands: A case study from Uganda. *Wetlands Ecology and Management, 22*(2), 199–213.

Climate as a Driver of Biodiversity Change on Land

Jyoti Rani, Sahil Singh Dhull, Deepak Beniwal, and Vibhuti Gulia

Abstract

At the end of twenty-first century climate change is a significant factor and one of the drivers of biodiversity change. Climate change has appeared as one of the most pressing worldwide challenge, intensely impacting different characteristics of our natural ecosystem. Among the impacts possessed by climate change on ecological concerns biodiversity loss and change in pattern of biodiversity gathered noteworthy attention from environmentalist, conservationists, scientists, and policymakers alike. Nowadays global warming is already disturbing ecosystem and species around the world, predominantly the most susceptible ecosystems like polar, coral reefs, and mountains ecosystems. Furthermore, climate change has also altered different ecosystem services on which people's livelihoods depend, like soil fertility and rainfall which are vital to agricultural production. Besides, dropping greenhouse gas productions is imperious to mitigate the extent of climate change and its impacts on biodiversity. Interdisciplinary research by uniting ecology, climatology and conservation biology is indispensable to address the interactions between climate change and biodiversity. Long-term monitoring, robust modelling, and adaptive management practices will be key to enhancing the resilience of ecosystems and preserving biodiversity. By using traditional knowledge and engaging local in conservation strategies in ecosystem conservation will protect biodiversity. This chapter highlights the urgent need of action to adapt and mitigate to climate change to protect biodiversity on land.

J. Rani (✉) · S. S. Dhull · D. Beniwal · V. Gulia
Department of Botany, Chaudhary Devi Lal University, Sirsa, Haryana, India
e-mail: drjyotisangwan@cdlu.ac.in

Keywords

Climate · Biodiversity change · Ecosystem · Land pattern · Traditional knowledge

1 Introduction

Since the start of the twentieth century, the climate has been experiencing a gradual rise in temperature, which has been linked to the release of greenhouse gases. This connection has been recognized since 1975. The fundamental catalyst for climate change is the consistent rise in greenhouse gas (GHG) emissions resulting from human activity. In previous decades, the atmospheric temperature was regulated within an acceptable range. This was achieved by reflecting a portion of the sunlight that reaches the Earth back into space, while the remaining sunlight converted into heat, which kept the Earth and the atmosphere at a suitable temperature for the survival of living forms (Shivanna, 2022). Rising atmospheric temperature has significant implications for biodiversity, ecosystems, and human welfare. The primary indicators of climate change are the extensive historical records documenting CO_2 concentrations, global temperatures, and weather patterns. In previous decades, it was feasible to accurately forecast yearly weather patterns, including the onset and conclusion of monsoon rains. As a result, farmers were able to strategically schedule the planting of their crops to align with the prevailing weather conditions. Currently, the weather pattern is undergoing frequent annual changes, resulting in significant losses for farmers. Likewise, the precise amount of rainfall that occurs each year and the specific areas that experience excessive or scarce rainfall can no longer be accurately foreseen. Several regions formerly characterized by little precipitation are now seeing significantly increased rainfall, whereas locations

J. A. Parray (ed.), *Soil and Land Use Change*, Sustainable Landscape Planning and Natural Resources Management,
https://doi.org/10.1007/978-3-031-93075-1_13

historically known for excessive rainfall are experiencing a decrease in the amount of precipitation. Likewise, the duration and amount of snowfall in temperate zones have also become more unpredictable (Hausfather et al., 2020). Climate change presents significant risks to biodiversity. While some degree of climate shifts is necessary for the survival and functioning of ecosystems, a sudden and abrupt change is harmful to biodiversity. The future is predicted to see an intensification of biodiversity loss due to climate change. Numerous species may be unable of adjusting to the swiftly changing, likely unfavorable environment, hence facing the risk of extinction. Climate change may have already caused the extinction of many species in recent times. In the past century, many species have shifted towards the higher elevations and poles, and this trend is expected to continue. Local communities are becoming more fragmented and include a greater number of species that are suited to warmer environments. Phenological changes in populations, such as alterations in breeding cycles or delays in growth peaks, are causing species relationships to become disconnected. Phenological changes in plants might be triggering incompatibility between plant and pollinators. This might possibly result in the abolition of both the plants and the pollinator, which would have significant implications for the composition of these mutually beneficial networks. Climate change aggravates the consequences of habitat loss and fragmentation on biodiversity by making fragmented biological populations more vulnerable to extinction risks. The loss and fragmentation of habitat may greatly impede the mobility of species and their capacity to adapt to climate change by following optimal climatic conditions. Even landscapes that are relatively undamaged are in danger, especially in cases where there is limited variation in the terrain, which requires animals to possibly travel long distances in order to find optimal climatic conditions. Spatial heterogeneity may provide certain species with a protective buffer against the effect; however, the extent of this buffering will differ depending on the location. Population reactions to severe climatic measures like flood, fire, and other natural calamities may be influenced by various habitat characters like area, quality and heterogeneity. Relations between land-cover and climate change are likely to be universal occurrences and have the ability to primarily impact the extent and distribution of decreases in biodiversity (Mantyka-Pringle et al., 2015).

Islands, especially small low-lying reef islands, are geographically limited ecosystems. Species inhabiting tiny islands have restricted options for adapting to changing environmental circumstances via range shifts due to the limited land area available. However, the species inhabiting islands must possess the ability to endure frequent fluctuations in their surroundings. Island species may have challenges in adapting to climate change owing to many factors, such as their restricted geographic ranges, limited genetic diversity, tiny populations that colonize new areas, lower species diversity, and inadequate adaptations to evade predators. The direst effect of climate change might be the extinction of whole species if individuals are unable to adapt or adjust (Taylor & Kumar, 2016). Examining methodologies and tactics to alleviate a shifting climate offers avenues to decrease the depletion of biodiversity and opportunities for sustaining nourishment in the future. It is crucial to demonstrate the worldwide effects of Climate Change on biodiversity using concrete scientific facts. Furthermore, the results of this study have significance in identifying, measuring, and examining the present and anticipated effects of Climate Change on variations in biodiversity. This information is valuable for policymakers, researchers, and interested organizations. The current scientific facts and information on global phenomena and the interconnection and interdependence of its components are valuable for developing global solutions and gaining insights. This work seeks to enhance comprehension and facilitate communication among scientific communities, researchers, and national and international policymakers about the connections between Climate Change and biodiversity (Muluneh, 2021). Furthermore, its primary objective is to provide current scientific knowledge about the effects of Climate Change on biodiversity alteration and the corresponding measures for adaptation. Hence, addressing Climate Change and biodiversity loss need the collective attention and intellect of every country and community.

2 Climatic Factors Affecting Biodiversity

Climatic conditions have a crucial role in determining the distribution, abundance, and survival of species, making them one of the most significant forces affecting biodiversity. The variability in climate, which may occur naturally or as a result of human activities, can have significant effects on ecosystems and the creatures that inhabit them. The following are the primary climate elements that impact biodiversity:

2.1 Temperature Fluctuation

Variation of temperature affects biodiversity in many ways. Increase or decrease in temperature alters the metabolic rates of species by changing their energy requirements and normal operations. Species frequently shift their geographic ranges to keep them in new temperature optima since this allows them to occupy areas that will offer them favorable living conditions. Through heat stress, high temperatures lead to stress on plants and animals, with indications on growth and sometimes the ability of an organism to reproduce. The stress weakens immunity to the organism concerned, thus predisposing it to disease and reducing its competitive advantage (Williams & Newbold, 2019). Other factors include interference with some biological events such as breeding, migration, or flowering. Plants that bloom seasonally will, if they shift to a different time, no longer coincide with important pollinators and thus can affect reproduction for the plant and for food supply for other species. Yet another potential impact seeking invasive species includes changes in the distribution of some species in a growing season. Warmer temperatures may open up countries or areas for the colonization of non-native species that previously could not survive and therefore compete with and displace native species. This may result in native biodiversity loss and alter ecosystem functioning. The changes in temperature can have an effect on aquatic conditions by the provided oxygen in the water, affecting fish and other aquatic animals. Warm waters induct the process of coral bleaching, wherein the symbiotic algae are expelled by coral tissues, further leading to the degradation of coral reefs—a critical habitat for many marine species (Prakash, 2021).

2.2 Precipitation Patterns

Since this affects water availability—an important factor in plant growth and therefore soil moisture—it will affect biodiversity. Progressive habitat alteration, by changing precipitation patterns, may render the habitat uninhabitable for its original species. The drought—a long period of inadequate rainfall—decreases food and water sources for animal populations in extreme numbers. More rainfall and an increase in flooding could also result in habitat destruction and erosion. A lot of water can leach down the top-soil and lower the overall fertility of the land; it may change the structure of entire ecosystems. Some species might be put at enormous risk of extinction under wetter conditions or in case of stable environmental losses (Wieczynski et al., 2018). The change in precipitation pattern may affect flow and volume in rivers, lakes, and other freshwater ecosystems, upon which many species rely for stable water conditions. Changes in precipitation manifest themselves by a change in the timing or availability of some resources. For example, the uncertainty of rainfall can affect the time of flowering or fruiting of some plants, affecting the species that depend on such plants for food. The wetland ecosystems are sensitive to the changing of the water level; hence, the changes can shift the ecological balance and, in effect, affect the diversity of flora and fauna of these wetland ecosystems (Trew & Maclean, 2021).

2.3 Extreme Weather Events

Extreme weather events such as hurricanes, flooding, and wildfires can instantly wipe out an ecosystem. In fact, most of the time such events normally cause great habitat loss associated with a decline and eventually a loss in biodiversity. The low mobility of species or the ability to adjust easily to becoming the cause of them developing more susceptibility to such events. Recovery from such extreme weather events is usually slow, and in many cases, ecosystems may never recover. The loss of a key species is likely to be realized as an indirect effect on other species in the same food web, changes in ecosystem processes, and a decrease in biodiversity. For example, wildfire guts a forest; it may result in a subsequent loss of habitat for very many species, thereby affecting the whole ecosystem. Not only may such an unfortunate event result in a short-term change but also long-term effects. For instance, hurricanes are able to change the coastal landscapes and therefore may affect marine and terrestrial species. Flooding may cause changes in plant communities related to sedimentation and change in soil chemistry. In case of wildfires, some fire-adapted species will simply increase, while the non-adapted ones will decrease; in that way, the species composition and ecosystem structure will be changed (Shivanna, 2022).

3 Mechanisms of Climate Impact on Biodiversity (Fig. 1)

3.1 Habitat Fragmentation and Loss

Habitat fragmentation and loss are increased due to climate change, which is the substantial threat to biodiversity. With rising temperatures across the globe, ecosystems continue changing, and there is no other option but to migrate to new areas in search of better living conditions. Habitat fragmentation is normally the result, whereby populations are usually isolated in certain habitats that often lead to loss of genetic diversity, hence increasing the chances of extinction. Also, human-induced habitat fragmentation is characterized by an outcome of activities such as agriculture, urbanization, and construction of infrastructures

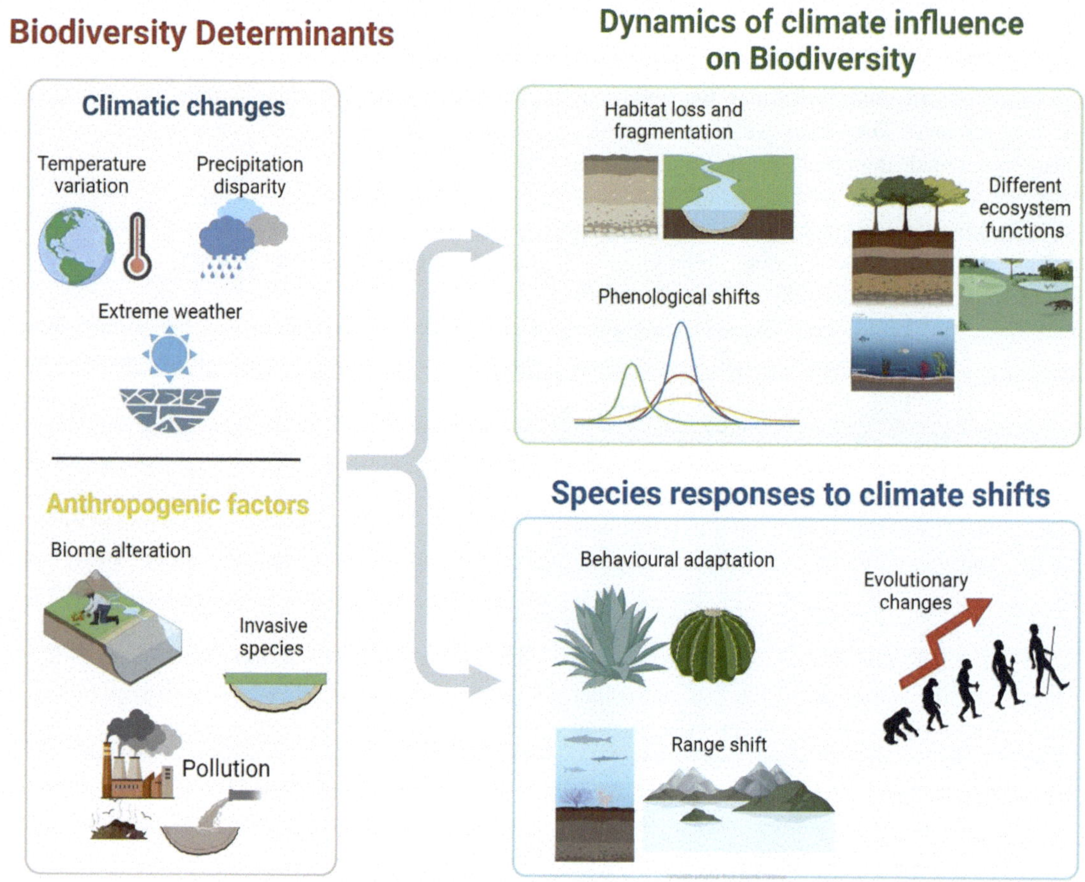

Fig. 1 Interplay of biodiversity determinants and climate influence on ecosystems

breaking these habitats into pieces with barriers characterized by limited species movement and resource accessibility ~~(Mendes & Prevedello, 2020)~~. Habitat fragmentation leads to population decrease and increase in isolation, by which the population becomes vulnerable to stochastic events of extinction due to extreme weather, outbreaks, or bottlenecks in genetics. Thus, loss of biodiversity through fragmentation is irreversible since such an isolated population cannot recolonize an area where conditions improve later on. Fragmented landscapes introduce edge effects; for instance, altered conditions at habitat edges may have adverse effects on a species, making interior habitat less suitable (He et al., 2019).

3.2 The Various Functions of Ecosystems

With climatic change, many of the natural self-sustaining processes may be influenced, as in cycling of nutrients, purification of water, and pollination. Changes in temperature and precipitation patterns could modify the productivity of ecosystems and their dependent species. For instance, changing temperature will shift the time of plant growth and its extent, therefore altering the plants' reproduction. This will further have its impacts on herbivores that feed on the plants and on the predators that feed on the herbivores. One such effect of this cascading process may alter species composition, which could bring down certain functions and hence reduce biodiversity. Furthermore, extreme weather or climatic events—such as hurricanes, droughts, and floods—are capable of directly influencing functions through the physical destruction of habitats and reducing the abundance of some essential resources. Effects of local extinction of keystone species may be especially severe. For instance, too many herbivores can cause overgrazing effects of vegetation after the loss of top predators (Pugnaire et al., 2019).

3.3 Phenological Shifts

Phenology refers to the timing of events like flowering and breeding in organisms and migration. Climate change has thus been illustrated to alter these phenological events, and for some species, it results in a shift that is out of phase with their environment. For instance, earlier warming in spring may cause plants to flower before the emergence of their pollinators, hence disturbing the plant–pollinator interactions. Also, advances in dates of insect emergence alter the species of birds whose calendars couple their breeding season to insect availability, apparently reducing reproductive success and survival rates of many of the involved species, thus reducing biodiversity. Apart from plant–pollinator and predator–prey interactions, competitive relationships between species may be impacted by phenological shifts. This is because, should some species change their phenology very fast to the changing conditions, then they will have added advantage over those who do not, potentially changing the composition of communities and excluding less adaptable species. This will thus ultimately affect the very services that human societies depend on, including crop pollination and pest control, again showing the breadth of implications due to climate-induced changes in biodiversity (Zettlemoyer & Peterson, 2021).

4 Response of Species to Climate Change

Species exhibit diverse reactions to climate change, which are influenced by their biology, ecology, and the unique environmental changes they encounter. These factors include alterations in the geographic range and distribution of species, modifications in the timing of biological events, adjustments in genetic traits to suit new conditions, fluctuations in population sizes, and, in exceptional cases, and the disappearance of a species from the Earth. Responses of species to climate change:

4.1 Behavioral Adaptations

Environmental changes impact behaviors mostly by affecting factors that regulate the fitness like opportunity to find food, avoiding predation, or having access to mates, parental care, and social interactions with both conspecifics and heterospecifics. In the greatest number of cases, climate changes possess several negative effects on fitness, like habitat destruction dropping survival and reproductive rate of a large number of species. This makes the individuals modify their behavior for survival maintain a high fitness. For a restricted number of species, environmental changes increase fitness, or new pathways to increasing fitness are opened up by behavioral modifications, often at the cost of other species. Species adjust their behaviors to better cope with new environmental pressures brought on by climate change. This can involve changes in migration patterns, where birds might migrate earlier or later in the year based on temperature shifts or food availability. Some species may alter their territorial behavior or mating rituals to adapt to new environmental cues. Additionally, changes in social structures or communication methods can occur; for instance, social insects like ants might adjust their foraging strategies or nest-building behaviors in response to changing temperatures and resource availability. These behavioral adaptations are crucial for maintaining survival and reproductive success as environmental conditions evolve (Wong & Candolin, 2014).

4.2 Evolutionary Changes

Evolutionary adaptations are more gradual and take longer timescales compared to behavioral adaptations. They involve genetic changes through the process of natural selection in populations. For example, plants can evolve traits such as drought-resistant root systems or changed leaf structure to cope with low water availability over a prolonged period. Similarly, changes in physiological traits that enhance an animal's survival capacity against new extremes of temperature extremes—improved heat-dissipating mechanism or changed metabolic rates—are examples of evolutionary adaptation. Such evolutionary changes can be realized through changed reproductive strategies, for example, by breeding seasons shifting to new climatic conditions or by changing the frequency of reproduction to get the highest possible survival of offspring under the new conditions. The evolution of these traits is usually preconditioned on high genetic variation within populations and may be determined by the rate of environmental change and species' ability to adapt (Aguirre-Liguori et al., 2021).

4.3 Range Shifts

As global temperatures rise, many species are shifting their ranges to find more suitable climates. For example, mountain-dwelling species might move up in elevation, while some marine species are migrating towards cooler, deeper waters. These shifts can impact ecosystems in complex ways, such as altering species interactions, predator–prey dynamics, and competition for resources. The introduction of species into new areas can sometimes lead to invasive behaviors, where non-native species out compete local species for resources. Barriers to movement, such as human infrastructure or unsuitable intermediate habitats, can restrict range shifts and lead to localized extinctions or population declines. The movement of species can also affect ecological balance, as new interactions are formed in these altered landscapes (Pinsky et al., 2020).

5 Interaction with Other Anthropogenic Factors

Nowadays biodiversity loss is among the top issues on the minds of intellectuals. These threats result due to environmental issues that are brought about by both natural and manmade causes. In actuality, human activities that impact biodiversity are serious environmental problems. These possess an undesirable impact on all living forms, not just humans. Biodiversity is under danger practically everywhere on Earth due to human activity and population expansion. Changes in land use, pollution, overuse of resources, and invasive species are some of the local hazards to species richness. One practical instrument for managing the environment and natural resources is biodiversity conservation (Prakash & Verma, 2022).

5.1 Land Use Change

Land use refers to the manner in which people use and manage land, together with its natural environment, to fulfil their requirements. Variation in land use pattern are driven by either natural factors or manmade factors both operating on different scales and magnitudes. The researchers have compounded the impact of both natural and human factors and attributed natural factors attributed to causing land use changes are climate change, volcanic eruptions, wildfires, and pests. Events of infestation, flooding, and drought, and their underlying anthropogenic drivers of change in land use are categorized either as direct impacts of human activities or as indirect impacts of human activities. Anthropogenic factors are to a large degree acknowledged as key drivers of land use changes, therefore about one-third to half of the world's land surface changes have been attributed to human activities, like overexploitation, conversions to agricultural land, overgrazing, urbanization, and development (Table 1) (Munthali et al., 2019).

Agriculture plays a significant role in Land Use and has had a substantial influence on the kinds of vegetation found in wetlands. This is mostly due to the expansion of crop areas, the increase in livestock numbers, and the use of agricultural inputs. Land use activities such as cultivation, animal grazing, and settlements are the primary practices in the vicinity of wetlands in the Lake Victoria basin. This issue is worsened by current policies that promote the draining of these areas for extensive agriculture by private corporations or individuals. These will alter local climates—urban heat islands—and act as barriers to species movement, potentially leading to distribution shifts of species and a reduction in biodiversity. Climate change can enhance these effects by making urban environments more hostile for many species to survive.

5.2 Pollution

Depending on the actual pollutant load and the soil's sensitivity to a particular pollutant, different impacts of air

Table 1 Impact of land use on biodiversity change

Land-use type	Description	Primary impact on biodiversity	Examples of affected species/habitats	Conservation strategies	References
Agriculture	Conservation of natural landscapes to cropland or pasture	Habitat loss, soil degradation, pesticide Impact	Grassland birds, pollinators, soil organisms	Sustainable farming, agroforestry, crop diversification	(Keprate et al., 2024)
Urbanization	Expansion of cities and infrastructure	Fragmentation pollution, invasive species	Wetland, urban wildlife, amphibians	Urban planning, green spaces, wildlife corridors	(Li et al., 2022)
Deforestation	Clearing of forests for timber or agriculture	Habitat degradation, loss of tree species	Tropical Forests, primates, forests birds	Reforestation, protected areas, sustainable logging	(Betts et al., 2017; Bousfield et al., 2023)
Mining	Extraction of minerals and fossil fuels	Soil and water contamination, habitat destruction	Aquatic ecosystem, endemic plant species	Land Reclamation, pollution control, restoration ecology	(Sonter et al., 2020)
Intensive livestock	Large scale animal farming operations	Overgrazing, pollution, methane emissions	Grasslands, rivers, carnivores	Rotational grazing, organic farming, riparian buffers	(Sigsgaard & Balmes, 2017)
Tourism	Development of tourism infrastructure	Disturbance, habitat alteration, resource depletion	Costal ecosystems, mountain habitats	Eco-tourism, environmental impact assessment, zoning	(Baloch et al., 2023)
Plantation forestry	Monoculture tree plantation for timber/pulp production	Loss of biodiversity, soil degradation, water use	Native forests, Soil invertebrates	Mixed species plantation, buffer zone, certification	(Liu et al., 2018)

pollutant deposition are experienced by different soil types. Sulphur, nitrogen, and heavy metals are released into the atmosphere by the burning of fossil fuels in processes like transportation, industry, and power generation, as well as through no combustion processes like waste management and fertilizer application in agriculture. When these pollutants are deposited as dry or wet deposition, they are transported off-site and may even traverse international borders. Deposition affects remote areas, and it is worse in forests and at higher elevations because of slower wind and more precipitation, respectively. Deposition exceeds the "critical load" that a certain soil can support, which negatively affects the soil's structure and functionality. Nitrogen oxide and sulfur dioxide are two pollutants that can change plant communities and alter soil chemistry, hence modifying species composition and interaction. Climate change may interact with these effects, modifying pollutant dispersion and deposition and further exacerbating the impact on ecosystems. Nutrient pollution in agricultural and industrial runoff disturbs aquatic ecosystems and species dependent upon them. It can thus influence the dispersal and magnitude of such pollution and the resistance of the affected ecosystems (Jaureguiberry et al., 2022). Chemical Contaminants Pesticides, heavy metals, and other chemicals can have direct, bioaccumulating, toxic impacts on wildlife and plants, hence changing the function of ecosystems. Climate change can exacerbate these impacts by changing contaminant distribution and altering the capacity of species to cope with pollutants.

5.3 Invasive Species

This could be facilitated by climate change, particularly as a result of changes in temperature and precipitation patterns, and additional new habitats might become suitable. In general, invasive species are superior competitors to native species, hence diminishing native biodiversity, while in some cases causing functional change in native ecosystems. The potential for invasive species to disrupt fundamental ecological interactions and processes of communities includes such basic interactions as predation, competition, and mutualism. Climate change can modify these interactions—for example, through time and intensity of ecological processes, increasing the impacts of invasive species. Invasive species alter habitats and ecosystem functions—for example, interfering with nutrient cycling or having an impact on soil structure. Climate change can go on to influence these modifications through its impacts on the growth and spread of invasive species and the resilience of ecosystems to those changes (Jaureguiberry et al., 2022).

6 Predictive Models and Future Projections

Ecologists, conservation experts in biology, and resource administrators have a significant challenge in comprehending, forecasting, and minimizing the effects of global change on biodiversity in the present century. Projections indicate that anthropogenic climate change is expected to result in extinctions, both alone and in conjunction with other endangering factors. However, the extent of its overall influence is concerning due to the wide range of estimations. To gain a better understanding of the impact of climate change and other environmental shifts on biodiversity in the future, it is necessary to adopt a more evidence-based approach in global change ecology. This will enable us to design effective conservation strategies that can mitigate the loss of biodiversity caused by climate change. Recent research at the intersection of palaeoecology, climatology, population genetics, and quantitative ecology is providing valuable insights into how species and populations have historically responded to global environmental changes (Fordham et al., 2016).

6.1 Climate-Biodiversity Model

Climate-biodiversity models are sophisticated instruments that combine elements of climate modelling with biodiversity research to forecast the potential effects of climate change on species, ecosystems, and total biodiversity. Various models are used, each with a specific emphasis on distinct elements of the correlation between climate and biodiversity. Here are some key types:

Species distribution models (SDMs): SDMs are widely used in the fields of ecology, evolution, and conservation. The emergence of readily available software packages and the growing accessibility of digital geoinformation have significantly facilitated the implementation of SDMs in the last ten years. This has dramatically expanded their use in informing conservation and management efforts, as well as in assessing the effects of global change. Nevertheless, models must be tailored to meet specific requirements, taking into account all crucial elements of their creation and use. A series of process that follows to the five fundamental SDM modelling steps are conceptualization/overview, data, model fitting, assessment, and prediction (Zurell et al., 2020). Species distribution models function based on a fundamental concept that utilizes essential data inputs, including occurrence and environmental characteristics, to estimate the locations where a species is likely to be found. The fundamental input in such models is geospatial data, which may be generated by documenting the locations where a species has been seen. This data may be sourced from historical field surveys, museum archives, or citizen science databases. In addition, environmental data within the framework of SDMs include factors such as temperature, precipitation, soil composition, elevation, and vegetation density. All of these factors contribute to our comprehension of the habitat preferences of the species under investigation. The first step involves establishing the statistical correlation between the specie's occurrence data and the environmental factors. These interactions demonstrate the particular circumstances in which a species may thrive, and hence, the species distribution model encompasses a "climate envelope" or ecological niche. By having a well-defined niche, the model can more accurately predict the possible range of the species across various geographic regions. Projections often center on current or future circumstances, with the latter derived mostly from climate models to anticipate how the range would be modified in response to climate change. In summary, this technique equips SDMs with the necessary tools to accurately forecast the outcomes of changes in specie's distribution, hence providing valuable insights for conservation and management purposes (Elith & Leathwick, 2009).

SDMs play a crucial role in ecology and conservation due to their several important functions and applications. Initially, they are used to forecast future distributions under various climate change scenarios. This function serves as the foundation of conservation planning, as it allows for the identification of places that might continue to be suitable for species survival in the face of changing environmental circumstances, as well as locations that may need specific conservation actions. Furthermore, SDMs play a crucial role in assessing the impacts of climate change on biodiversity. These models may be used to simulate the potential changes in specie's ranges, allowing for the identification of vulnerable species at risk of extinction or in need of specific habitat preservation. This enables proactive conservation efforts to be implemented. Additionally, they are valuable in the control of invasive species, since they allow for the prediction of regions where the species may spread and enable early intervention to avoid or mitigate the invaders' negative effects on the native ecosystems. Furthermore, SDMs are used in evaluating habitat suitability to determine the optimal sites for species reintroduction or habitat restoration. This ensures that conservation efforts are concentrated on regions with the highest probability of supporting the long-term survival and persistence of species (McSHEA, 2014).

Dynamic vegetation models (DVMs): DVMs use a process-based methodology to replicate the demographic characteristics of plant populations. They have been utilized to investigate many aspects such as disturbances, plant succession, community composition, and the provision of ecosystem services in the context of climate change scenarios. Although they have the capacity to accomplish so, they have seldom been used for the detailed examination of species range dynamics. Employing a dynamic model that incorporates many interacting mechanisms offers various benefits for investigating plant range alterations. The presence or absence of a species at a specific location is determined by various factors, including interactions with lower-level processes and higher-level constraints. These factors include the physiological response to the environment, limitations on dispersal, interactions with other organisms, and historical contingencies that may arise due to management practices. It is crucial to include many processes and their interactions in order to accurately represent non-linear and non-additive connections. Furthermore, a process-based approach allows for adaptability in creating innovative interactions under different environmental circumstances. For instance, individual species react to climatic changes autonomously, meaning that species move on their own rather than as a whole community. Furthermore, the flexible characteristics of DVMs enable us to investigate inquiries about the timing and mechanisms of range alterations. DVMs have the ability to include long-term ecological processes like as succession, as well as delays induced by dispersion restriction and biotic interactions. Ultimately, it is probable that distinct mechanisms play a significant role in the forefront compared to the rear end of a migratory species. When studying the extinction rates of trees in specific areas, it is crucial to take into account their lifespan, flexibility, and tolerance. These factors have a significant role in determining the pace at which other species progress in those areas (Snell et al., 2014).

Ecosystem Process Models: The term "ecosystem model" encompasses a diverse range of depictions of ecosystems, all of which have the common purpose of portraying the interactions between different components and/or processes within the ecosystem, such as species, functional groupings, and disturbances. These ecosystem models are mostly illustrative and aid in foundation for more complex ecological models. For instance qualitative models that use signed digraphs may be employed to investigate pulse-press perturbations in network sizes that range from small to substantial.

Ecosystem interactions are represented by Bayesian belief networks as a series of probabilistic events, showing how variations in the probability of one ecosystem element influencing the probability of another ecological element changing. These models may be shown as a graphical structure or a sequence of probabilities. Fuzzy cognitive maps are being widely used to provide semi-quantitative predictions about the effects of changes, such as reintroductions or invasive species management, on ecological networks. Ecosystem models that use a dynamical systems theory approach employ a deterministic methodology to forecast the temporal evolution of ecosystems.

These models usually rely on Lotka–Volterra equations, or similar ones, and may need a significant amount of data, particularly if the model is intricate. These methodologies have been used to forecast the comprehensive consequences of species reintroductions, seed distribution on islands, and the results of invasive species management, and may have wider implications (Geary et al., 2020).

Population viability studies (PVAs): PVAs play a significant role in the development of conservation theory, policy, and management. The majority of PVAs concentrate on individual species within a particular environment and aim to tackle a specific issue. Other things that population viability analyses can do for conservation include the following: they help in understanding the different factors bearing most influence on population dynamics and viability, provide an estimate of extinction risk, identify key habitats, compare ecological and socioeconomic trade-offs, set priorities in management options, convey the message of crisis in conservation to various stakeholders, and identify gaps in research. Many PVA research lack organization, which makes them difficult to comprehend, evaluate, replicate, or use for generating broad conclusions. An assessment comparing published PVAs and existing guidelines revealed several shortcomings. Firstly, model selection was rarely justified. Secondly, important parameters were neglected or described vaguely. Thirdly, limited details were provided on parameter ranges, sensitivity analysis, and scenarios. Lastly, the results were often reported inconsistently, making it difficult to repeat the study or compare it with others. While there are numerous guidelines available for designing and implementing dependable PVAs, and standards for documenting and communicating ecological models in general, there is a noticeable absence of organized guidelines specifically tailored to designing, applying, and communicating PVAs that consider their varied structures and contents (Peer et al., 2013).

Life cycle assessment (LCA) Models with Climate and Biodiversity Focus: offers a structured approach for businesses to review the consequences of their product and supply chain choices. Recent advancements in methodology have expanded the scope of environmental effects to include biodiversity and ecosystem services. Despite its usefulness, the use of LCA is still restricted when it comes to assessing land-use change. Currently, the techniques used to estimate land-use change rely on extrapolating historical trends at the national level using linear models. Furthermore, the terrain inside countries or ecoregions is uniform, with spatial variation occurring exclusively within these specific groupings. Furthermore, the analysis fails to consider landscape configuration, such as habitat fragmentation, and context, such as proximity to landscape features like watercourses. However, it is important to acknowledge that local spatial heterogeneity and landscape configuration play a crucial role in determining the effects on biodiversity and various ecosystem services. Ultimately, the use of predictive framing, which is usually limited to consequential life cycle assessment (LCA) focused on modifications or marginal impacts, would also be advantageous for attributional LCA that focuses on accounting or non-marginal impacts. This would allow for the anticipation of overall changes in corporate and sectoral decision-making. Life Cycle Assessment (LCA) is a valuable approach for pinpointing areas of significant environmental impact within supply chains, examining the environmental effectiveness of product design decisions, and aiding in the establishment of environmental laws and regulations by both public and private entities. Given the significant role of habitat loss and fragmentation in causing a decline in global biodiversity, it is crucial to consider the effects on biodiversity that arise from land use practices. This is a key focus in the development of LCA. Recently, researchers in the field of life cycle assessment (LCA) have looked to the literature on ecology and conservation for ideas on how to create models for assessing the effect of life cycles on biodiversity. This includes using meta-analyses and Species-Area Relationship (SAR) models to measure the loss of biodiversity (Curran et al., 2016).

6.2 Scenarios of Biodiversity Change

Biodiversity is essential for the provision of ecosystem functions and services. However, it is now experiencing a significant global decrease as a result of environmental changes caused by human activities. Effective governance and proactive management of biodiversity need the development of credible scenarios that anticipate the potential changes that may occur in response to future environmental circumstances. In recent decades, various ecological models have been created to generate biodiversity forecasts. These models take into account and predict the effects of two key factors that shape the future of biodiversity: (1) the expected pattern, speed, and severity of environmental changes, and (2) the capacity of living organisms to adapt to these changes through ecological and evolutionary processes. While several modelling techniques consider the mechanisms that impact species distribution, abundance, community structure, and ecosystem function, only a limited number of research investigate the adaptive capacity of species. Several research articles have forecasted the potential impacts of anticipated climate change on biodiversity. Future climate change is forecasted to cause changes in species ranges, either in terms of latitude or altitude, which will have significant impacts on ecological communities. Additionally, it is expected to heighten the likelihood of species extinction and diminish the efficacy of conservation zones (Titeux et al., 2016). Biodiversity change scenarios are predictions of how biodiversity may alter in the next decades based on various environmental, economic, and policy circumstances. Scenarios enable scientists, policymakers, and conservationists to comprehend the potential impacts of climate change, land use changes, pollution, and conservation measures on ecosystems and species in the future. Several typical situations for examining changes in biodiversity include:

Business-as-usual (BAU) Scenarios—Scenarios of biodiversity change under normal business operations (BAU): Projections refer to the predicted future condition of biodiversity based on existing trends, policies, and practices without any substantial changes. In this scenario, the current patterns of land use, resource extraction, pollution, and climate change persist without any significant alterations, and no new conservation initiatives or policy interventions are implemented. This scenario serves as a reference point for evaluating the possible effects of different methods aimed at maintaining biodiversity and fostering development. The reconfiguration of all landscapes is expected to result in continuous habitat degradation and fragmentation due to its main association with the conversion of land for agriculture, urbanization, and infrastructure development. This devastation leads to a significant negative impact on species with extensive geographical ranges. Climate change will worsen these effects by altering temperature and precipitation patterns, as well as sea levels, which will all have an influence on the distribution of species, their migratory patterns, and the overall functioning of ecosystems. The ongoing pollution resulting from industrial operations, agriculture, and waste management will persist in causing damage to ecosystems and exacerbating ocean acidification, eutrophication, and the build-up of plastics in the environment. The BAU scenario would result in an expedited decline in species, deterioration of ecosystems, less genetic variation, and significant economic and societal consequences. The deterioration of ecosystems and the death of living species sometimes lead to a cascading effect that negatively impacts other sectors. This example demonstrates the potential detrimental consequences for biodiversity if these patterns continue without substantial involvement in implementing more sustainable practices and policies to mitigate biodiversity loss and preserve ecosystems for the benefit of future generations (Panwar et al., 2023).

Sustainable Development Scenario: A Sustainable Development Scenario for biodiversity change entails the need for economic development to coexist with environmental conservation and social fairness, in order to achieve favorable results for both human societies and natural ecosystems. This scenario involves the simultaneous implementation of proactive measures to reduce environmental pressures, protect biodiversity, and restore damaged ecosystems, while also ensuring the sustainability of livelihoods and promoting social well-being. It prioritizes conservation as a central aspect of larger development goals and guarantees that economic operations are favorable to the long-term well-being of the Earth's natural systems. The objective will be accomplished by the implementation of sustainable land use and agricultural practices, the mitigation and adaptation to climate change, the regulation of pollution, the attainment of resource efficiency, and the expansion of protected areas. Biodiversity is included into economic planning via the use of incentives that promote sustainable activities. Additionally, local communities play an active role in conservation efforts, ensuring social fairness and the equitable distribution of benefits. The resulting ramifications include the stability and recuperation of species populations, enhanced provision of ecosystem services, increased resistance to environmental changes, and sustainable livelihoods and economic advantages. This would need the cooperation and coordination of nations in order to achieve the established goals on biodiversity, while also providing help to poor areas. A Sustainable progress Scenario may provide a promising approach for the future, where human progress and biodiversity conservation work together to mutually support and protect the planet's natural heritage, ensuring long-term prosperity and well-being for future generations (Tran et al., 2020).

Global Conservation Effort Scenario: The Global Conservation Effort Scenario envisions a future in which the international community has successfully collaborated to safeguard and restore biodiversity globally. This scenario entails a coordinated and proactive worldwide effort to address the biodiversity catastrophe. It involves enhanced collaboration among governments, NGOs, companies, and local communities to implement comprehensive conservation initiatives. Therefore, its purpose is to ensure the protection of ecosystems, prevent the extinction of species, and maintain ecological processes necessary for the continuation of life on Earth. Contrary to the prevailing tendencies, this situation is motivated by the shared acknowledgement of the crucial significance of biodiversity for worldwide sustainability and human welfare. The overarching objectives are to significantly expand the coverage of protected areas on both land and in the water, rehabilitate extensively damaged ecosystems, strengthen the legal and regulatory frameworks, and incorporate biodiversity considerations into all levels of government and across all sectors of society. Undoubtedly, global collaboration and funding will be essential for this endeavor, namely ensuring sufficient financial resources for initiatives in poor countries via many channels and methods, while also prioritizing the empowerment of local populations to lead conservation efforts. Furthermore, it is essential to complement all of this with public awareness and education initiatives aimed at fostering a culture of conservation. The Scenario is characterized by the prevention of mass extinction, the preservation of ecological services, the development of resistance to climate change, and the addition of economic benefits and societal well-being. Ultimately, it assigns people the role of caretakers and emphasizes the responsibility linked to global collaboration and ethical obligation to save our precious world for future generations (Nicholson et al., 2019).

Climate Mitigation and Adaptation Scenario: The Climate Mitigation and Adaptation Scenario provides a comprehensive outlook on the future, focusing on intensified measures to decrease greenhouse gas emissions in the atmosphere, as well as the adaptation of human and natural systems to the unavoidable consequences of climate change. The large-scale measures in this scenario pertain to reducing emissions by implementing renewable energy on a wide scale, improving energy efficiency, adopting sustainable land use practices, and engaging in reforestation. In addition to these efforts, adaption measures are being undertaken to enhance the resilience of ecosystems and human communities in response to climate change. The measures include the preservation and rehabilitation of vital habitats, facilitation of species movement to more favorable surroundings, and establishment of infrastructure and agricultural systems capable of withstanding harsh weather events and changing climatic conditions.

The scenario has a very advantageous impact on biodiversity as it reduces the extent of climate change, so allowing ecosystems and species more time and resources to adapt to the new conditions. Primarily, this scenario mitigates the severity of global warming by addressing the significant impact of climate change on habitat disruption, which includes coral bleaching, loss of polar ice, and the transformation of verdant grasslands and forests into deserts. Specifically, implementing climate adaption measures with conservation efforts that focus on specific targets, such as protecting habitats and restoring ecosystems, significantly increase the likelihood of stabilizing, if not restoring, biodiversity. Species facing the risk of extinction due to fast environmental changes have a higher chance of surviving and thriving if their habitats are conserved or restored. Additionally, the development of migration corridors can aid in their relocation to more favorable conditions.

In this scenario, the probability of the continued existence of the provision of fundamental services from these ecosystems, like as carbon storage, water filtering, and pollination, is optimized. These services are very significant for both human well-being and economic stability. The synergistic method of mitigating and adapting to slow down the rate of biodiversity loss involves enhancing the resilience of natural systems, so assuring their continued functioning and support for life on Earth. The Climate Mitigation and Adaptation Scenario depicts a future trajectory that is somewhat positive, with proactive actions enhancing biodiversity results and fostering a more sustainable and resilient global environment (Howarth & Robinson, 2024).

6.3 Uncertainties and Limitations

Although there is a noticeable need for scenarios that provide different biodiversity outcomes based on future emissions, there is currently a lack of comprehensive study and communication of the uncertainties involved. In this study, we simulated the worldwide distribution of around 11,500 species of amphibians, birds, and mammals. We then predicted their adaptation to different climates in the years 2050 and 2070, while manipulating the input data. This study investigates the uncertainties that arise from the selection of species distribution models (SDMs), dispersion methods, global circulation models (GCMs), and representative concentration routes (RCPs).

Species-level uncertainty: At the level of each species, the impact of species distribution models (SDM), general circulation models (GCM), and representative concentration pathways (RCP) varied depending on the specific sensitivity measurement and dispersion scenario used. When assuming restricted dispersion, the selection of species distribution models (SDMs) results in a deviation in the change of appropriate habitats that is more than 10 times greater than the deviance caused by the selection of the other two components. Put simply, the subjective choice of a particular SDM has a significant impact on the ultimate outcome, in contrast to the selection of GCMs and RCPs. When examining simply the decline in climatic suitability under the assumption of no dispersion, it is noteworthy that RCPs have the greatest influence on the variance, followed by SDMs and GCMs. The mismatch in species distribution models (SDMs) in estimating the future climatic adaptability of species is the reason for the observed difference in the explained deviation between the two dispersal assumptions. SDMs often exhibit strong agreement when forecasting the present distribution of species. Put simply, the variation in model output is not just determined by the quality of the SDMs, but also by their internal structure. Complex models, such as random forest, utilize different combinations of features to make predictions about geographical conditions. However, these predictions may vary greatly when applied to future conditions, especially if the models have been calibrated using specific conditions. However, when taking into account the potential loss of climatic habitat that is now appropriate (as assumed by the no dispersion assumption), the impact of Representative Concentration Pathways (RCPs) on the level of uncertainty in projections becomes more significant as the extent of the species' range rises. SDMs and RCPs have a significant role in explaining uncertainty for the rarest species, whereas for species with extensive ranges, the impact of RCPs on the overall variation in future climatic suitability is the greatest (Thuiller et al., 2019).

Model-based uncertainty: The fundamental objective and difficulty in conservation is to assess and enhance the accuracy of biodiversity projections in order to provide more effective guidance for policy choices. Implementing a thorough approach to assess and minimize the uncertainty of model outputs in comparison to observed measurements and numerous models will enhance the reliability of biodiversity forecasts. Our proposed technique combines biodiversity models with new remote sensing and in-situ data streams to assess and minimize uncertainty, with the aim of enhancing policy-relevant predictions on biodiversity. Models are crucial instruments for integrating ecological and socio-economic information in order to enhance comprehension and governance of social-ecological systems. Difficulties embrace, between other things, the accurate depiction of feedback loops between the socio-economic and ecological subsystems, uncertainty, and human behavior. Models rely on environmental input, such as temperature and precipitation data, which may be affected by measurement mistakes and have uncertainties in predicting future climate. This fluctuation will affect the dependability of forecasts for these variables. Many models need the use of simplified assumptions when it comes to species distributions, climatic effects, and ecological interactions. These assumptions may sometimes fail to adequately capture the intricacies of real-world ecosystems, which might result in erroneous findings (Drechsler, 2020).

Biological and ecological Uncertainties: Data from throughout the globe indicates that human climate change is impacting biological systems at many levels, ranging from ecosystems to genes. Temperature and precipitation fluctuations, along with the resulting changes in winter harshness, plant life cycles, drought and animal conditions, spread of alien species and predation, abundance and disease, may impact ungulates either directly or indirectly. The responses of ungulates to climatic fluctuations and changes, however, are not uniform and vary depending on the species and geographical location. Currently, there is a lack of comprehensive and uniform reporting of model-related uncertainty in ecological and evolutionary investigations. Three primary obstacles that have contributed to gaps in uncertainty consideration include a preoccupation with uncertainty linked to parameters, unclear uncertainty measurements, and insufficient acknowledgement of uncertainty transmission. Nevertheless, these disparities may be eliminated. We suggest that enhancing uncertainty reporting in the domains of ecology and evolution may be achieved by using established statistical methods more extensively and by incorporating best practices from other scientific disciplines. Our suggestions include giving more thought to uncertainties related to input data and model structure, establishing particular criteria for uncertainty in field-specific procedures and reporting, and enhancing uncertainty propagation by using hierarchical models (Simmonds et al., 2024).

Socio-Economic Factors: To effectively manage ecosystems and preserve biodiversity, it is necessary to take into account not only ecological and evolutionary factors, but also the social and economic aspects that impact the management goals. In this article, we provide a number of theoretical models that focus on the economic and social factors of human civilization that are intimately linked to the management of ecosystems. The first model examines the economic choices about the use of the shared fishing area at Jeju Island, Korea, made by traditional divers and visitors. Land-use change is a major cause of biodiversity loss, although accurately predicting it is challenging due to its dependence on several socio-economic variables. Typically, predictive models make the assumption that data extrapolation will result in consistent patterns (Lee & Iwasa, 2014).

Biological data uncertainty: Species distribution data is often used to evaluate the susceptibility of species to climate and land use changes, as well as to enhance the effectiveness of species conservation initiatives. Through geographic forecasts, species distribution models (SDMs) provide this kind of information; however, the underlying species occurrence data amount and quality determine how accurate and precise the projections are. The availability of data to support species distribution models is increasing quickly, with sources such as citizen science and new sensing technologies contributing significantly. However, it is important to note that these data have limitations in advancing our understanding of global species distribution due to sampling biases. A primary objective of contemporary species distribution modelling (SDM) techniques is to address this limitation by integrating several forms of species occurrence data, hence enhancing spatial predictions.

7 Conservation Strategies and Policy Implications

7.1 Climate Adaptation Strategies and Conservation Planning

Considering these rapidly changing climate trends, addressing the issues of adaptation strategies to safeguard the ever-more-endangered biodiversity is central to realizing any gains. The Intergovernmental Panel on Climate Change Intergovernmental Panel on Climate Change (IPCC), 2023 explains that "adaptation is the adjustment process to actual or expected climate and its effects". Adaptation strategies are not in any way homogenized and have to be tailored according to the exact contexts of every ecosystem. In what follows, some of the climate adaptation measures to counter that changing landscape are examined (Fig. 2).

7.1.1 Land Protection and Management

More protection has to come in place for enhanced protection for increased diversity in the habitat. Additionally, organic and regenerative farming among other sustainable ways which will help in supporting some aspects of biodiversity and contributing to climate change mitigation will also have place in the wider land-use strategies. This biodiversity conservation, in addition to adaptation to climate change, includes widening the scope of protected areas and ensuring that there is good representation and replication of a variety of ecosystems and species across other various protected areas for greater general sustainability of ecosystems (Mawdsley et al., 2009).

Management of protected areas also means that some effects of climate change can be ultimately countered by the planting of trees, building of sea barriers, and the use of controlled fires to prevent the devastation from uncontrolled wildfires. Treatment of the smaller areas parks and reserves–are easier to control and some specific restoration measures have been tested for some types of ecosystems; still, such efforts may be expensive and limited in potential. The surrounding environment is also important because changes outside protected areas are likely to have significant impacts on what may happen within it. In thinking location and slope, natural sites added today and restoration sites are designed to attenuate the effects of climate change on issues like the increase of the level of the sea. Different elevations of protected spaces can help species to adapt in different climatic condition changes, ensuring space for movement species (Smith et al., 2020). Using mixtures of more hardy plants species can make a restoration more effective, but with the potential for introducing an invasive species problem. Even when it's known that some projects cannot be sustainable in the long-term, they can still provide temporary refuges for species of concern until alternative and more successful strategies are identified (Felton et al., 2016).

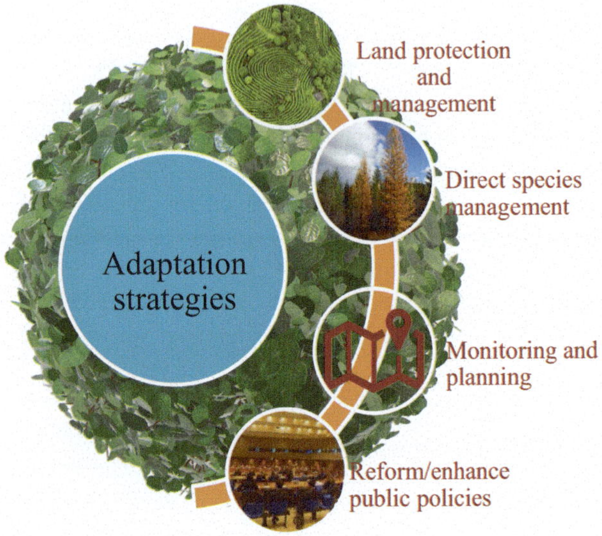

Fig. 2 Climate adaptation strategies

7.1.2 Solutions to Regulate Species Directly

The key contribution of work on rare and climate-threatened species is that every "endangered" species cannot be protected due to the limit in funds for conservation. The means are not enough with regard to the available funding for the conservation of these species. Those operations that would secure each species at stake are, for this reason, difficult to conduct. It will be hard to mobilize new financial support to implement effective conservation actions for each climate-affected endangered species. Creating captive populations is a last-ditch effort for species on the brink of extinction due to climate change. And while there are some techniques, like seed storage and propagation, to accrue plants in captivity, maintaining these sorts of programs is likely intensive and, in the long term, unworkable for most species because of the potential for climate change to resoundingly reshape the nature of most of their habitats (Rawat & Agarwal, 2015).

7.1.3 Approaches Related to Monitoring and Planning

Monitoring systems enhance informed decision-making by managers through the provision of relevant information, especially during periods that are characterized by accelerated environmental change, in which these systems function. Dynamic landscape conservation plans include fixed and dynamic spatial elements that are areas of change in land use. They provide guidelines for the management of target species, genetic resources, and ecosystems within the planning areas. Desirable future conditions are defined for each component, under projected species and ecosystem distribution changes.

Table 2 International agreements and policy frameworks and their key points

S. no	Agreement/framework	Key points
1	Kunming-Montreal global biodiversity framework	– Protect 30% of land and water key for biodiversity by 2030
		– Raise $200 billion by 2030 for biodiversity
		– Phase out or reform subsidies which could provide an additional $500 billion for nature
		– Increase financial support to poor countries from the current target of $20 billion annually by 2025 to $30 billion by 2030
2	UN Framework Convention on Climate Change (UNFCCC) and UN Convention on Biological Diversity (CBD)	– Recognize the interconnectedness between climate change and biodiversity

Intermediate objective conditions can be set that would assist the movement of the species from a current to a future condition. Such plans target to adapt the biodiversity to climate change and are consistent with other regional management plans (Mawdsley et al., 2009).

7.1.4 Vision Related to Law and Policy

There is a need for the constant readjustment of public policies concerning the management of wildlife and biodiversity. This involves allowing the implementation of instruments such as legislation, regulation, and private-sector standards. Furthermore, laws and policies concerning wildlife, natural resources, and biodiversity must be revisited in order to make them compatible with what managers require in meeting the challenge of climate change effects. This can be attained by the employment of legislation, regulations, private sector standards, policies and codes, complemented by the action of compliance and enforcement. Amend public policies in the area, attainable by reviewing existing laws, regulations, and policies on natural resource management and amending those instruments to have uniformity relating to the situation that managers are facing associated with the impacts of climate change (Van asselt et al., 2008).

The adaptation strategies should be done while looking at biodiversity and not focusing only on human health and infrastructure since it is the provider of services such as pollination and water filtration. Integration of human and biodiversity needs in times of crisis hence becomes key brought about by climate change intervention strategies and should work in harmony and not in conflict.

7.2 International Agreements and Policy Frameworks

In December 2022, the Biodiversity Convention decided on the Kunming-Montreal Global Biodiversity Framework, which would replace the Aichi Biodiversity Targets that were set back in 2010 (Final Text of Kunming-Montreal Global Biodiversity Framework Available in All Languages, 2022). It is supposed to be an ambitious and effective post-2020 global biodiversity framework accompanied by specific goals and benchmarks that could put nature and humans on the right course. This framework will operate in concert with the Paris Climate Agreement and other global accords on forests, desertification, and oceans.

The two accords are the UN Framework Convention on Climate Change and the United Nations Convention on Biological Diversity. While the two treaties were created to address different challenges of climate change and biodiversity during the 1992 Rio Earth Summit, they also recognize their interconnectedness and harmony. Collectively, these frameworks provide Nature-Based Solutions (NBS) to protect natural ecosystems (e.g., coral reefs, mangroves) to combat climate change and safeguard biodiversity (Table 2) (Pettorelli et al., 2021).

8 Conclusion

Climate change is a significant factor that causes biodiversity and ecosystem functioning to change due to rising temperatures, changed rainfall patterns and recurrent severe weather events. These changes exert pressure on species to adapt, move, or face extinction, hence altering the makeup and functioning of ecosystems. While some species may demonstrate behavioral or evolutionary adaptation to changing environmental circumstances, others may lack the ability to do so, leading to a decline in biodiversity. The latter are exacerbated by the intricate interplay between climate change and other anthropogenic activities associated with changes in land use. Incorporating these effects into conservation efforts would need the use of climate models, safeguarding and rehabilitation of habitats, and the implementation of multidisciplinary methodologies at the intersection of ecology, climatology, and conservation biology. The urgent need to ensure the enduring health and stability of terrestrial ecosystems is increasingly imperative due to the simultaneous challenges of climate change and biodiversity loss. This

assumes that addressing these challenges would need coordinated efforts in conservation, mitigation, and adaptation techniques. Protecting biodiversity is essential for maintaining robust ecosystems and enhancing human resilience against the detrimental effects of climate change. Hence, it is imperative to take immediate, synchronized actions on a worldwide scale to prevent the detrimental effects of climate change and save the planet's abundant biodiversity.

References

Aguirre-Liguori, J. A., Ramírez-Barahona, S., & Gaut, B. S. (2021). The evolutionary genomics of species' responses to climate change. *Nature Ecology & Evolution, 5*(10), 1350–1360. https://doi.org/10.1038/s41559-021-01526-9

Baloch, Q. B., Shah, S. N., Iqbal, N., Sheeraz, M., Asadullah, M., Mahar, S., & Khan, A. U. (2023). Impact of tourism development upon environmental sustainability: A suggested framework for sustainable ecotourism. *Environmental Science and Pollution Research, 30*(3), 5917–5930. https://doi.org/10.1007/s11356-022-22496-w

Betts, M. G., Wolf, C., Ripple, W. J., Phalan, B., Millers, K. A., Duarte, A., Butchart, S. H. M., & Levi, T. (2017). Global forest loss disproportionately erodes biodiversity in intact landscapes. *Nature, 547*(7664), 441–444. https://doi.org/10.1038/nature23285

Bousfield, C. G., Lindenmayer, D. B., & Edwards, D. P. (2023). Substantial and increasing global losses of timber-producing forest due to wildfires. *Nature Geoscience, 16*(12), 1145–1150. https://doi.org/10.1038/s41561-023-01323-y

Curran, M., Maia de Souza, D., Antón, A., Teixeira, R. F. M., Michelsen, O., Vidal-Legaz, B., Sala, S., & Milà i Canals, L. (2016). How well does LCA model land use impacts on biodiversity?—A comparison with approaches from ecology and conservation. *Environmental Science & Technology, 50*(6), 2782–2795. https://doi.org/10.1021/acs.est.5b04681

Drechsler, M. (2020). Model-based integration of ecology and socio-economics for the management of biodiversity and ecosystem services: State of the art, diversity and current trends. *Environmental Modelling & Software, 134*, Article 104892. https://doi.org/10.1016/j.envsoft.2020.104892

Elith, J., & Leathwick, J. R. (2009). Species distribution models: Ecological explanation and prediction across space and time. *Annual Review of Ecology, Evolution, and Systematics, 40*(1), 677–697. https://doi.org/10.1146/annurev.ecolsys.110308.120159

Felton, A., Gustafsson, L., Roberge, J. M., Ranius, T., Hjältén, J., Rudolphi, J., Lindbladh, M., Weslien, J., Rist, L., Brunet, J., & Felton, A. M. (2016). How climate change adaptation and mitigation strategies can threaten or enhance the biodiversity of production forests: Insights from Sweden. In *Biological conservation* (vol. 194, pp. 11–20). Elsevier Ltd. https://doi.org/10.1016/j.biocon.2015.11.030

Final text of Kunming-Montreal Global Biodiversity Framework available in all languages. (2022). https://www.cbd.int/conferences/2021-2022/cop-15/documentsasdocument:CBD/COP/15/L25

Fordham, D. A., Akçakaya, H. R., Alroy, J., Saltré, F., Wigley, T. M. L., & Brook, B. W. (2016). Predicting and mitigating future biodiversity loss using long-term ecological proxies. In *Nature climate change* (vol. 6(10), pp. 909–916). Nature Publishing Group. https://doi.org/10.1038/nclimate3086

Geary, W. L., Bode, M., Doherty, T. S., Fulton, E. A., Nimmo, D. G., Tulloch, A. I. T., Tulloch, V. J. D., & Ritchie, E. G. (2020). A guide to ecosystem models and their environmental applications. *Nature Ecology & Evolution, 4*(11), 1459–1471. https://doi.org/10.1038/s41559-020-01298-8

Hausfather, Z., Drake, H. F., Abbott, T., & Schmidt, G. A. (2020). Evaluating the performance of past climate model projections. *Geophysical Research Letters, 47*(1). https://doi.org/10.1029/2019GL085378

He, X., Liang, J., Zeng, G., Yuan, Y., & Li, X. (2019). The effects of interaction between climate change and land-use/cover change on biodiversity-related ecosystem services. *Global Challenges, 3*(9). https://doi.org/10.1002/gch2.201800095

Howarth, C., & Robinson, E. J. Z. (2024). Effective climate action must integrate climate adaptation and mitigation. *Nature Climate Change, 14*(4), 300–301. https://doi.org/10.1038/s41558-024-01963-x

Intergovernmental Panel on Climate Change (IPCC). (2023). Glossary. In *Climate change 2022—Impacts, adaptation and vulnerability* (pp. 2897–2930). Cambridge University Press. https://doi.org/10.1017/9781009325844.029

Jaureguiberry, P., Titeux, N., Wiemers, M., Bowler, D. E., Coscieme, L., Golden, A. S., Guerra, C. A., Jacob, U., Takahashi, Y., Settele, J., Díaz, S., Molnár, Z., & Purvis, A. (2022). The direct drivers of recent global anthropogenic biodiversity loss. *Science Advances, 8*(45). https://doi.org/10.1126/sciadv.abm9982

Keprate, A., Bhardwaj, D. R., Sharma, P., Verma, K., Abbas, G., Sharma, V., Sharma, K., & Janju, S. (2024). *Climate resilient agroforestry systems for sustainable land use and livelihood* (pp. 141–161). https://doi.org/10.1007/978-3-031-63430-7_7

Lee, J. H., & Iwasa, Y. (2014). Modeling socio-economic aspects of ecosystem management and biodiversity conservation. *Population Ecology, 56*(1), 27–40. https://doi.org/10.1007/s10144-013-0423-0

Li, G., Fang, C., Li, Y., Wang, Z., Sun, S., He, S., Qi, W., Bao, C., Ma, H., Fan, Y., Feng, Y., & Liu, X. (2022). Global impacts of future urban expansion on terrestrial vertebrate diversity. *Nature Communications, 13*(1), 1628. https://doi.org/10.1038/s41467-022-29324-2

Liu, C. L. C., Kuchma, O., & Krutovsky, K. V. (2018). Mixed-species versus monocultures in plantation forestry: Development, benefits, ecosystem services and perspectives for the future. *Global Ecology and Conservation, 15*, Article e00419. https://doi.org/10.1016/j.gecco.2018.e00419

Mantyka-Pringle, C. S., Visconti, P., Di Marco, M., Martin, T. G., Rondinini, C., & Rhodes, J. R. (2015). Climate change modifies risk of global biodiversity loss due to land-cover change. *Biological Conservation, 187*, 103–111. https://doi.org/10.1016/j.biocon.2015.04.016

Mawdsley, J. R., O'Malley, R., & Ojima, D. S. (2009). A review of climate-change adaptation strategies for wildlife management and biodiversity conservation. In *Conservation biology* (vol. 23(5), pp. 1080–1089). https://doi.org/10.1111/j.1523-1739.2009.01264.x

McSHEA, W. J. (2014). What are the roles of species distribution models in conservation planning? *Environmental Conservation, 41*(2), 93–96. https://doi.org/10.1017/S0376892913000581

Muluneh, M. G. (2021). Impact of climate change on biodiversity and food security: A global perspective—a review article. *Agriculture & Food Security, 10*(1), 36. https://doi.org/10.1186/s40066-021-00318-5

Munthali, M. G., Davis, N., Adeola, A. M., Botai, J. O., Kamwi, J. M., Chisale, H. L., & Orimoogunje, O. O. (2019). Local perception of drivers of land-use and land-cover change dynamics across Dedza District, Central Malawi Region. *Sustainability, 11*(3), 832. https://doi.org/10.3390/su11030832

Nicholson, E., Fulton, E. A., Brooks, T. M., Blanchard, R., Leadley, P., Metzger, J. P., Mokany, K., Stevenson, S., Wintle, B. A., Woolley, S. N. C., Barnes, M., Watson, J. E. M., & Ferrier, S. (2019). Scenarios and models to support global conservation targets. *Trends in Ecology & Evolution, 34*(1), 57–68. https://doi.org/10.1016/j.tree.2018.10.006

Anwar, R., Ober, H., & Pinkse, J. (2023). The uncomfortable relationship between business and biodiversity: Advancing research on business strategies for biodiversity protection. *Business Strategy and the Environment, 32*(5), 2554–2566. https://doi.org/10.1002/bse.3139

Pe'Er, G., Matsinos, Y. G., Johst, K., Franz, K. W., Turlure, C., Radchuk, V., Malinowska, A. H., Curtis, J. M. R., Naujokaitis-Lewis, I., Wintle, B. A., & Henle, K. (2013). A protocol for better design, application, and communication of population viability analyses. *Conservation Biology, 27*(4), 644–656. https://doi.org/10.1111/cobi.12076

Pettorelli, N., Graham, N. A. J., Seddon, N., da Cunha, M., Bustamante, M., Lowton, M. J., Sutherland, W. J., Koldewey, H. J., Prentice, H. C., & Barlow, J. (2021). Time to integrate global climate change and biodiversity science-policy agendas. *Journal of Applied Ecology, 58*(11), 2384–2393. https://doi.org/10.1111/1365-2664.13985

Pinsky, M. L., Selden, R. L., & Kitchel, Z. J. (2020). Climate-driven shifts in marine species ranges: Scaling from organisms to communities. *Annual Review of Marine Science, 12*(1), 153–179. https://doi.org/10.1146/annurev-marine-010419-010916

Prakash, S. (2021). Impact of climate change on aquatic ecosystem and its biodiversity: An overview. *International Journal of Biological Innovations, 3*(02). https://doi.org/10.46505/ijbi.2021.3210

Prakash, S., & Verma, A. K. (2022). Anthropogenic activities and Biodiversity threats. *International Journal of Biological Innovations, 4*(1), 94–103. https://doi.org/10.46505/ijbi.2022.4110

Pugnaire, F. I., Morillo, J. A., Peñuelas, J., Reich, P. B., Bardgett, R. D., Gaxiola, A., Wardle, D. A., & Van Der Putten, W. H. (2019). Climate change effects on plant-soil feedbacks and consequences for biodiversity and functioning of terrestrial ecosystems. *Science Advances, 5*(11). https://doi.org/10.1126/sciadv.aaz1834

Rawat, U. S., & Agarwal, N. K. (2015). Biodiversity: Concept, threats and conservation. *Environment Conservation Journal, 16*(3), 19–28. https://doi.org/10.36953/ECJ.2015.16303

Shivanna, K. R. (2022). Climate change and its impact on biodiversity and human welfare. *Proceedings of the Indian National Science Academy, 88*(2), 160–171. https://doi.org/10.1007/s43538-022-00073-6

Sigsgaard, T., & Balmes, J. (2017). Environmental effects of intensive livestock farming. *American Journal of Respiratory and Critical Care Medicine, 196*(9), 1092–1093. https://doi.org/10.1164/rccm.201706-1075ED

Simmonds, E. G., Adjei, K. P., Cretois, B., Dickel, L., González-Gil, R., Laverick, J. H., Mandeville, C. P., Mandeville, E. G., Ovaskainen, O., Sicacha-Parada, J., Skarstein, E. S., & O'Hara, B. (2024). Recommendations for quantitative uncertainty consideration in ecology and evolution. *Trends in Ecology & Evolution, 39*(4), 328–337. https://doi.org/10.1016/j.tree.2023.10.012

Smith, P., Calvin, K., Nkem, J., Campbell, D., Cherubini, F., Grassi, G., Korotkov, V., Le Hoang, A., Lwasa, S., McElwee, P., Nkonya, E., Saigusa, N., Soussana, J., Taboada, M. A., Manning, F. C., Nampanzira, D., Arias-Navarro, C., Vizzarri, M., House, J., & Arneth, A. (2020). Which practices co-deliver food security, climate change mitigation and adaptation, and combat land degradation and desertification? *Global Change Biology, 26*(3), 1532–1575. https://doi.org/10.1111/gcb.14878

Snell, R. S., Huth, A., Nabel, J. E. M. S., Bocedi, G., Travis, J. M. J., Gravel, D., Bugmann, H., Gutiérrez, A. G., Hickler, T., Higgins, S. I., Reineking, B., Scherstjanoi, M., Zurbriggen, N., & Lischke, H. (2014). Using dynamic vegetation models to simulate plant range shifts. *Ecography, 37*(12), 1184–1197. https://doi.org/10.1111/ecog.00580

Sonter, L. J., Dade, M. C., Watson, J. E. M., & Valenta, R. K. (2020). Renewable energy production will exacerbate mining threats to biodiversity. *Nature Communications, 11*(1), 4174. https://doi.org/10.1038/s41467-020-17928-5

Taylor, S., & Kumar, L. (2016). Global climate change impacts on pacific islands terrestrial biodiversity: A review. *Tropical Conservation Science, 9*(1), 203–223. https://doi.org/10.1177/194008291600900111

Thuiller, W., Guéguen, M., Renaud, J., Karger, D. N., & Zimmermann, N. E. (2019). Uncertainty in ensembles of global biodiversity scenarios. *Nature Communications, 10*(1), 1446. https://doi.org/10.1038/s41467-019-09519-w

Titeux, N., Henle, K., Mihoub, J., Regos, A., Geijzendorffer, I. R., Cramer, W., Verburg, P. H., & Brotons, L. (2016). Biodiversity scenarios neglect future land-use changes. *Global Change Biology, 22*(7), 2505–2515. https://doi.org/10.1111/gcb.13272

Tran, T. C., Ban, N. C., & Bhattacharyya, J. (2020). A review of successes, challenges, and lessons from Indigenous protected and conserved areas. *Biological Conservation, 241*, Article 108271. https://doi.org/10.1016/j.biocon.2019.108271

Trew, B. T., & Maclean, I. M. D. (2021). Vulnerability of global biodiversity hotspots to climate change. *Global Ecology and Biogeography, 30*(4), 768–783. https://doi.org/10.1111/geb.13272

van Asselt, H., Sindico, F., & Mehling, M. A. (2008). Global Climate Change and the Fragmentation of International Law. *Law & Policy, 30*(4), 423–449. https://doi.org/10.1111/j.1467-9930.2008.00286.x

Wieczynski, D. J., Boyle, B., Buzzard, V., Duran, S. M., Henderson, A. N., Hulshof, C. M., Kerkhoff, A. J., McCarthy, M. C., Michaletz, S. T., Swenson, N. G., Asner, G. P., Bentley, L. P., Enquist, B. J., & Savage, V. M. (2018). Climate shapes and shifts functional biodiversity in forests worldwide. *Proceedings of the National Academy of Sciences, 116*(2), 587–592. https://doi.org/10.1073/pnas.1813723116

Williams, J. J., & Newbold, T. (2019). Local climatic changes affect biodiversity responses to land use: A review. *Diversity and Distributions, 26*(1), 76–92. https://doi.org/10.1111/ddi.12999

Wong, B. B., & Candolin, U. (2014). Behavioral responses to changing environments. *Behavioral Ecology, 26*(3), 665–673. https://doi.org/10.1093/beheco/aru183

Zettlemoyer, M. A., & Peterson, M. L. (2021) Does phenological plasticity help or hinder range shifts under climate change? *Frontiers in Ecology and Evolution, 9*. https://doi.org/10.3389/fevo.2021.689192

Zurell, D., Franklin, J., König, C., Bouchet, P. J., Dormann, C. F., Elith, J., Fandos, G., Feng, X., Guillera-Arroita, G., Guisan, A., Lahoz-Monfort, J. J., Leitão, P. J., Park, D. S., Peterson, A. T., Rapacciuolo, G., Schmatz, D. R., Schröder, B., Serra-Diaz, J. M., Thuiller, W., & Merow, C. (2020). A standard protocol for reporting species distribution models. *Ecography, 43*(9), 1261–1277. https://doi.org/10.1111/ecog.04960

The manufacturer's authorised representative in the EU is Springer
Nature Customer Service Centre GmbH, Europaplatz 3, 69115 Heidelberg,
Germany. If you have any concerns regarding our products, please
contact ProductSafety@springernature.com

Printed and bound by CPI Group (UK) Ltd, Croydon, CR0 4YY
05/10/2026
02228164-0002